Tore Hägglund
Process Control in Practice

Also of interest

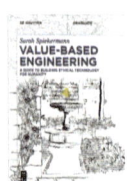

Value-Based Engineering
A Guide to Building Ethical Technology for Humanity
Sarah Spiekermann, 2023
ISBN 978-3-11-079336-9, e-ISBN (PDF) 978-3-11-079338-3

Technology Development
Lessons from Industrial Chemistry and Process Science
Ron Stites, 2022
ISBN 978-3-11-045171-9, e-ISBN (PDF) 978-3-11-045163-4

Basic Process Engineering Control
Paul Serban Agachi , Mircea Vasile Cristea,
Emmanuel Pax Makhura, 2020
ISBN 978-3-11-064789-1, e-ISBN (PDF) 978-3-11-064793-8

Empathic Entrepreneurial Engineering
The Missing Ingredient
David Fernandez Rivas, 2022
ISBN 978-3-11-074662-4, e-ISBN (PDF) 978-3-11-074682-2

Process Systems Engineering
For a Smooth Energy Transition
Edited by Edwin Zondervan, 2022
ISBN 978-3-11-070498-3, e-ISBN (PDF) 978-3-11-070520-1

Tore Hägglund

Process Control
in Practice

Translated by Margret Bauer

DE GRUYTER

Title of the Swedish original: *Praktisk processreglering*

Author
Prof. Tore Hägglund
Västgötavägen 9
SE-222 25 Lund
Sweden
tore.hagglund@control.lth.se

ISBN 978-3-11-110372-3
e-ISBN (PDF) 978-3-11-110495-9
e-ISBN (EPUB) 978-3-11-110746-2

Library of Congress Control Number: 2023939209

Bibliographic information published by the Deutsche Nationalbibliothek
The Deutsche Nationalbibliothek lists this publication in the Deutsche Nationalbibliografie;
detailed bibliographic data are available on the Internet at http://dnb.dnb.de.

Preface

This book is aimed at people who work in the process industry and come into contact with control technology, such as operators, technicians and engineers. It is also suitable for use in schools and higher education institutions where control technology is included in the syllabus. The book includes an exercise section with solutions. This gives the reader an opportunity to actively work with the methods addressed and also makes the book more suitable for use in schools and in education. It is written assuming only mathematical knowledge at school level. The content here covers the practical parts of control technology currently used in the process industry and it contains many examples from this industry. The technical development in the field of control technology has developed quickly in recent decades. This does not only apply to controllers and control systems, but also to other tools for engineers and operators. This technical development has led to major efficiencies and higher quality in the process industry

Technological developments and intensifying competition have resulted in a reduction in personnel. The rapid technological changes have also meant that many older people have left their roles and have been replaced by younger ones. These generational changes have often been a prerequisite for the industry to be able to take advantage of new computer-based techniques.

However, there is a danger with these generational changes. There is a risk that the process knowledge and experience found in the older generation is not passed on to the younger generation. The development means that new demands are placed on those who are responsible for instrumentation and operation of industrial processes. More advanced equipment offers greater opportunities than before, but for it to be able to be used efficiently also requires a knowledge of how it works, its opportunities and its limitations. The hope is that this book will help to convey parts of this knowledge.

This risk is exacerbated as we urgently need to respond to the threats resulting from our changing climate. The process industry is a large contributor to carbon emissions and the main user of natural resources. It is vital that we ensure efficient process operation. The first building stone in achieving this is good process control.

This is the first English translation of the original book 'Praktisk processreglering' published in Swedish by Studentlitteratur. There were four editions of the Swedish book: in 1990, 1997, 2008 and 2019. Over the years, the content of the book became known to colleagues and collaborators in industry and academia who asked for an English edition.

Large parts of the material in the book were used in practical control technology courses for the process industry, arranged primarily by ABB (formerly NAF Controls, SattControl Instruments and Alfa Laval Automation), SUM AB and in recent years Gustaf Fagerberg AB. A large part of the material and the gradual development of the book is the result of the knowledge and inspiration I got from course participants

https://doi.org/10.1515/9783111104959-202

and course leaders over the years and I am very grateful for the contact I had with the Swedish process industry. I also want to thank other friends and colleagues, both in the academic world and in industry, for stimulating discussions as well as contributions to the book. I would especially like to thank Karl Johan Åström, who always shared my interest and my research in the field, José Luis Guzmán for a very fruitful research collaboration, and Margret Bauer for not only translating the book but also suggesting several important improvements of the content. Finally, I would also like to thank Leif Andersson for invaluable help with the typographic design of the book and with the use of LaTeX.

Lund, March 2023   Tore Hägglund

A Note on the Translation

During my first year of my three-year appointment as Lise-Meitner professor at the department of Automatic Control at Lund University, Tore gave me a copy of the fourth edition of his Swedish book 'Praktisk processreglering' that had then just come out. Flipping through the pages I was immediately intrigued because of the clear and beautiful images, but could not understand most of it because of my limited command of the Swedish language. Talking to former ABB colleagues, who are fluent in Swedish, I found that I am not the first who has seen the necessity for a translation. After some discussions with Tore and the Swedish publisher we decided that it was time to start the international version of the book.

In front of you is the result. There is a myth that no one has ever read a book on control engineering cover to cover. This book challenges this myth. Not only is there hardly any math, Tore's writing is also highly readable and contains many insights into the practical side of process control that have not previously been recorded in a textbook. It also is short. This is because Tore has very carefully selected content, problems and solutions. For example, there are currently hundreds if not thousands published rules for PID tuning. In this book, four rules are carefully selected, discussed and compared.

With hindsight I wish that I had read this book much earlier in my career. Like many control engineers, I trained as an electrical engineer and was instantly attracted to the beautiful mathematical equations describing the world of control engineering. Once I started working in the process automation industry I was looking for applications of root locus, Nyquist, Bode and Nichols plot. Alas — there were none to be found! Only when I read Tore's book did I fully grasp why this is and I hope you will, too. In contrast, you will need (almost) everything written here if you actually have to control an industrial production process.

Over the decades, there have always been control engineers who have predicted the demise of the PID controller and regularly suggested novel, superior approaches. Needless to say, none of these approaches have actually succeeded in replacing PID control. Despite that, there are control journals that no longer publish research that is related to PID control. This is troublesome because reality is very different: industrial plants run on PID control, yet PID controllers are inherently poorly tuned. Processes require unnecessary human intervention if the control is not implemented practically. Until all controllers are well tuned and feedforward and ratio control are sufficiently exploited, there remain many unsolved research questions.

I have encountered MPC (model predictive control) experts who were called in to improve the performance of a production process but walked away after the initial assessment — simply because the base layer, comprising the PID controllers, was poorly tuned. MPC relies on well-tuned and performing base layer control. I have heard even

https://doi.org/10.1515/9783111104959-203

more control engineers in industry admitting to never using the D-part in the PID-controller. After reading this book, these situations should be a thing of the past.

The Swedish fourth edition of this book is flawless. Tore's writing is clear and the sentence structure in Swedish is simple. This all helps to make the complex topic of process control more accessible. I am a control engineer, versed in English, German and now some Swedish control terminology but not a professional translator. I hope I have done his writing justice.

Hamburg, March 2023   Margret Bauer

Contents

Preface —— V

A Note on the Translation —— VII

1   Introduction —— 1
1.1   Industrial Process Automation —— 1
1.2   Single Control Loop —— 2
1.3   Organisation of this Book —— 3

2   Process Types and Step Response Analysis —— 5
2.1   Introduction —— 5
2.2   Step Response Experiment —— 5
2.3   Common Process Types —— 8
2.4   Step Response Analysis —— 12
2.5   Other Analysis Methods —— 19

3   PID-Controllers —— 20
3.1   Introduction —— 20
3.2   PID-Controller Structure —— 20
3.3   Selection of Controller Types —— 31
3.4   PID-Controller Parameters —— 34
3.5   Identification of PID-Controllers —— 38
3.6   Practical Modifications of PID-Controllers —— 43

4   PID-Controller Tuning —— 52
4.1   Introduction —— 52
4.2   The Control Task —— 52
4.3   Manual Tuning —— 54
4.4   Step Response Methods —— 57
4.5   Self-Oscillation Methods —— 70
4.6   Automatic Tuning —— 77

5   Nonlinear Processes —— 81
5.1   Introduction —— 81
5.2   Nonlinear Valves —— 81
5.3   Friction —— 83
5.4   Backlash —— 88
5.5   Nonlinear Sensors —— 89
5.6   pH-Control —— 91

5.7	Asymmetrical Processes —— 92
5.8	Gain Scheduling —— 92
5.9	Adaptive Control —— 97

6	**Control Strategies** —— **101**
6.1	Introduction —— 101
6.2	Filters —— 101
6.3	Selector Control —— 108
6.4	Cascade Control —— 110
6.5	Feedforward —— 117
6.6	Mid-Range Control —— 126
6.7	Split-Range Control —— 132
6.8	Dead Time Compensation —— 135
6.9	Ratio Control —— 142
6.10	Decoupling —— 150
6.11	Buffer Control —— 154

Exercises —— 157
2	Process Types and Step Response Analysis —— 157
3	PID-Controllers —— 163
4	PID-Controller Tuning —— 165
5	Nonlinear Processes —— 168
6	Control Strategies —— 174

Solutions to the Exercises —— 178
2	Process Types and Step Response Analysis —— 178
3	PID-Controllers —— 184
4	PID-Controller Tuning —— 186
5	Nonlinear Processes —— 189
6	Control Strategies —— 191

Bibliography —— 193

Index —— 195

1 Introduction

The title of this book is *Process Control in Practice*. Most of the content is not limited to the process industry, but is useful in other industries where control technology is also used. However, most references will be made to the problems and needs of the process industry.

1.1 Industrial Process Automation

Most process industries have developed over a long period of time, in several cases for more than a century. Initially, these industries were run largely manually, which meant that people were out in the facilities and ensured that temperatures, pressures, flows, levels and other process variables were maintained within reasonable limits by manually adjusting valves, pumps and other control devices. Gradually, these industries became more automated, which meant that people were replaced by automation equipment and controllers. Staff nowadays rarely work inside the production facilities, but monitor the process from a control room. The number of employees in the process industry has also declined sharply in recent decades. This means that fewer and fewer people are monitoring larger and larger process sections. At the same time, the processes have become increasingly more sophisticated which has resulted in higher production yield and higher quality of the products.

The relatively slow and gradual increase of automation in the process industry has resulted in decentralised control strategies. In paper mills, for example, there are thousands of controllers that maintain process variables at desired values. These simple control loops are in turn interconnected in an ingenious network ensuring that the overall goal of the production is achieved. This network has been developed over a long time by many people and is based on in-depth knowledge of process operations and process characteristics.

A prerequisite for the development of the process automation industry is that the equipment used in automation has developed further. In the nineteen sixties, people first started using computers in the process industry. Prior to that, all controls were handled with analogue technology, usually with pneumatic devices. Towards the end of the nineteen seventies, computer-based controllers and control systems had a serious break-through in the process industry. The introduction of computers meant completely new possibilities for control technology solutions and monitoring of the processes.

It is interesting to note that despite the advent of computers in the process industries the processes today are largely controlled with the same control strategies that were employed before computers arrived. This has surprised many, but the reason is that the solutions developed by generations of people who lived close by the processes

https://doi.org/10.1515/9783111104959-001

are based on so much knowledge that they are difficult to surpass. Of course, there are cases where, for example, new optimisation methods have provided improved performance, but these usually work at a higher level where they determine setpoints for the existing underlying base level control described here. In this book we review the comparatively simple methods of base level control.

1.2 Single Control Loop

In the decentralised control strategies used in the process industry the single control loop is the basic and most important building block. A modern industrial process can comprise hundreds or thousands of individual control loops that all work together to manage the production.

Figure 1.1 shows a block diagram describing an individual control loop. The loop consists of two parts: the process and the controller. The process represents a process section, the output of which you want to control. This output can be the level in a tank, the flow or the pressure in a line, the temperature in an oven or the concentration in a flow.

The process connects the controller output to the process variable. In the figure, these are denoted by u and y, respectively. These designations have become standard in the control literature, but in product manuals you often see other designations. The process variable y (PV), sometimes also called measurement value, is the quantity we want to control and is in most cases the signal from a sensor. In order to be able to control the process variable, we must be able to influence the process. We do this by varying the controller output u, which often is the signal to a valve or a pump. The controller output (OP) is also sometimes referred to as the control variable or control signal.

The objective of the controller is that the process variable y should follow a setpoint. The setpoint (SP) is also called reference signal and is denoted by r in the figure. In some control applications, you want the process variable to follow a frequently changing setpoint. In process control, the setpoint usually does not to change often. Rather, outside disturbances make it difficult to keep the measurement value close to the setpoint.

The control loop in Figure 1.1 shows how to solve the control problem using feedback. In feedback control, the measurement signal y is compared to the setpoint r.

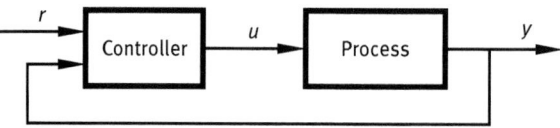

Fig. 1.1: A single control loop.

The comparison determines the value of the controller output u. The comparison and determination of the controller output is performed by the controller. The controller is the second important component in a control loop.

1.3 Organisation of this Book

The book first studies the single control loop, that is, the feedback loop of a process and a controller. This is done by reviewing various ways to describe the properties of the process, the structure of the controller and its tuning. Thereafter follows a discussion of the largest source of problems in control loops, namely, nonlinearities. The last chapter moves beyond single loop control by investigating what happens if several loops work together in one implementation. In other words, we start with a single control loop and work towards larger implementations and control strategies.

Process understanding is a prerequisite for successfully controlling the process. Chapter 2 describes different methods for investigating and describing the dynamic properties of processes. The most common way to determine the dynamic properties of processes in the process industry is to use step response experiments. The chapter describes different ways of analysing and describing the dynamics of processes with the help of such experiments. These descriptions form the basis of several of the methods to tune the controllers as described later in the book. This chapter also includes a classification of different types of processes that are commonly found in the process industries.

The PID-controller is by far the most common controller in the process industry. Chapter 3 describes the PID-controller in detail. First, the structure of the PID-controller is explained. Incidentally, the different parts of the PID-controller are a natural extension of the most basic type of controller — the on/off controller. The description of the structure of the PID-controller also provides natural explanations of how the controller's parameters affect the control performance. Different suppliers have chosen to implement the PID-controller in different ways. These variants are also discussed in this chapter.

Chapter 4 deals with different ways of tuning PID-controllers, both simple manual methods and methods based on the step response experiments described in Chapter 2. Methods based on examining the dynamics of the process by tuning the control loop so that it starts oscillating are also discussed. The chapter concludes with a description of the principles of automatic controller tuning.

Chapter 5 focuses on different types of nonlinearities that occur in control loops and describes how to deal with them. Nonlinearities often occur in valves in the form of backlash and friction. Even well maintained valves often have a nonlinear characteristic. Nonlinearities also often occur in sensors, and the process itself can be nonlinear. Gain scheduling is a systematic and relatively simple way to deal with many types of nonlinearities and is detailed in this chapter. The chapter concludes with a description of adaptive methods that can also be used to handle nonlinear processes.

Chapter 6 describes the most common control strategies used to connect single control loops. When instrumenting the controller, information is sent between different functions via signals. These signals may often need to be filtered so that they do not contain too much noise and other incorrect information. Different types of filters are described in Section 6.2. Selectors are logical components that are used to switch between different control strategies. This is described in Section 6.3. Cascade control, as described in Section 6.4, is a way of connecting two controllers and thereby having a controller structure that corresponds to a more advanced controller than the PID-controller. Feedforward is another way to improve the performance of the PID-controller. It is based on the principle that you should compensate for measurable disturbances even before they impact on the process variable. Feedforward is described in Section 6.5. Sometimes two control signals can be used to control a process variable. Mid-range and split-range control are two methods for this and they are described in Sections 6.6 and 6.7. Processes with long dead times require special types of controllers for high performance requirements. Principles for dead time compensation are reviewed in Section 6.8. Ratio control is a special combination of two controllers used to keep the ratio between two measurements constant. Ratio-control is described in Section 6.9. There are cases where two control loops interfere with each other in an undesirable way. In such cases, you can decouple the interaction to solve this problem. The principle of decoupling is explained in Section 6.10. The chapter ends with a description of buffer control in Section 6.11, which is a way of avoiding that different process sections interfere with each other.

Chapters 5 and 6 contain many examples of control problems taken from the process industry.

The book ends with an exercise section with solutions. The exercises are organised in the same way as the chapters in the book, so that you can easily find exercises that are related to each chapter. The exercise part provides an opportunity for a deeper understanding of the concepts covered in the book. Many of the exercises are based on industrial examples.

2 Process Types and Step Response Analysis

2.1 Introduction

The control task of a single control loop as described in Figure 1.1 is to determine the controller output u so that the process variable y follows the setpoint r. However, different processes react differently to changes in the controller output. Some processes react quickly, others slowly. Some processes react a lot, while others react very little. We call these properties the dynamics of the process. The different dynamics of different processes must be taken into account when designing a controller. This is done by tuning the controller parameters so that they are adapted to the process dynamics.

To achieve well performing control you must therefore first identify the dynamic properties of the process. This can be done in many different ways. For single loop control in the process industry, it is usually sufficient to conduct simple experiments. The most common method is the step response method. It is based on making an instantaneous change in the controller output from one value to another, referred to as a step change, and record the response in the process variable to the step change, the step response.

This step response provides information about the dynamics of the process. There are many tuning methods for PID-controllers based on step response experiments. Some of these are reviewed in Chapter 4. The analysis of the step response also provides information on where the biggest control challenges lie. Examples of control challenges are long process dead times, that is, the process does not react for a long time after the actuator has moved, or if valve stiction or other nonlinearities affect the process. Some strategies for appropriately modifying the controller to ensure good control for these situations are described in more detail in Chapters 5 and 6.

The step response analysis can also be used to study the characteristics of the closed control loop. Instead of putting the controller into manual mode and studying the connection between process input and output, it is possible to leave the controller in automatic mode and study the relationship between a setpoint change and the reaction in the process variable.

2.2 Step Response Experiment

Figure 2.1 shows a step response experiment of a process. In step response analysis, you put the controller in manual control, wait until the process variable is stationary and then suddenly change the controller output and thus the actuator, a valve or a pump, to a new value. The process dynamics can be determined by studying how this change in the controller output affects the process variable. There are several mea-

https://doi.org/10.1515/9783111104959-002

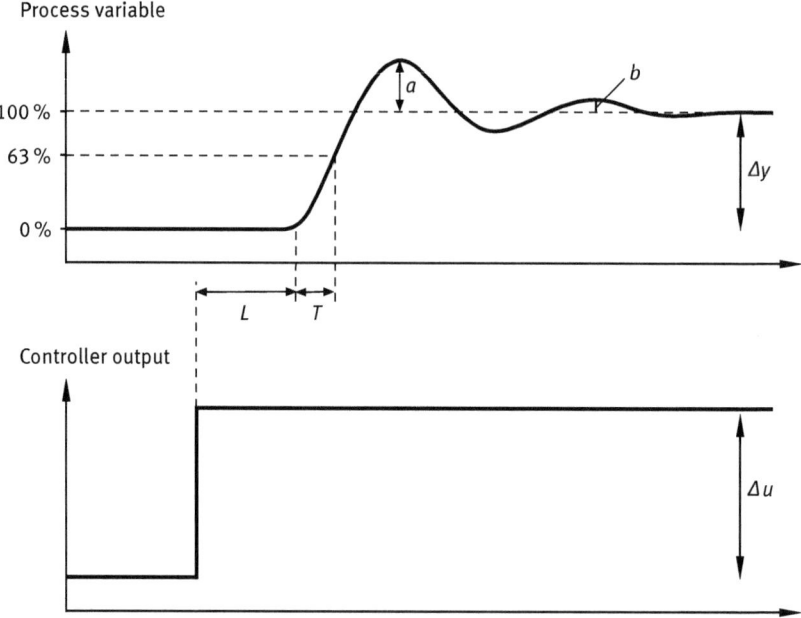

Fig. 2.1: Step response experiments indicating the most important parameters.

sures that describe the dynamic properties of the process and that are visible in the step response. Some of these can be seen in Figure 2.1.

Dead time L defines the time between changing the controller output and observing a reaction in the process variable. The dead time is also called the time delay.

Time constant T is a measure of how quickly the process variable responds to a step change after the dead time has elapsed. The time constant is defined as the time it takes for the process to reach 63 % of its final value. Note that the dead time is not included in this time. The reason why one has chosen the seemingly arbitrary number of 63 % has its basis in signal processing theory. The time constant is sometimes also referred to as the lag time or lag.

Steady-state gain $K_p = \Delta y/\Delta u$ indicates how much the process variable has changed when it has reached its final value in relation to the controller output change. It is also often called the process gain.

Overshoot $a/\Delta y$ [%] is the amount with which the process variable initially exceeds its target. It is usually stated as a percentage of the process variable change. In the process industry, the overshoot is almost always zero in uncontrolled processes.

Damping b/a is a measure of how quickly an oscillation decreases. Normally, processes in the process industry are well damped. In step response analysis we will therefore get zero damping in most cases. The damping is, however, a popular parameter for specifying the performance of the feedback controller.

Units and Measurement Ranges

It is important to think about working with the right quantities and units when describing the dynamics of a process. If the description is to be the basis for tuning controllers, it is important that the units are aligned with the controller units. The integral time and derivative time of the PID-controller, which will be introduced in Chapter 3, are indicated in some cases in seconds and in other cases in minutes. Dead time L and time constant T must have the same units as the controller.

The controller gain is usually based on normalised signals, which means that both the measurement value and the controller output are signals that vary between 0 % and 100 %. This is done even if the process variable can be presented in physical quantities. Since the gain of the controller is based on normalised signals, the process gain K_p must be normalised too, if you want to use it to tune the controller.

The controller output is almost always specified in percent, which means that the controller output range is $OP_{range} = 100\% - 0\% = 100\%$. As stated, the process variable is often presented in physical quantities and the process variable range PV_{range} is then given by the process variable range specified in the controller. Note that this does not have to be the same as the sensor's measuring range, although it usually is. Sometimes it can be difficult to find the process variable range in an existing control system configuration. Section 3.5 describes an experimental method for finding out which process variable range is specified in the controller. The following example shows which calculations may be needed to obtain the normalised process gain.

Example 2.1.   Controller output and process variable range

Assume that a step response experiment has been performed on a temperature controlled process where the controller output is specified in percent and where the process variable is specified in °C. The process variable range is 0 °C to 1000 °C. This means that $OP_{range} = 100\%$ and $PV_{range} = 1000\,°C$.

In the step response experiment, the controller output was changed from $u_1 = 30\%$ to $u_2 = 40\%$, that is, the controller output change was $\Delta u = 40 - 30 = 10\%$. This step change caused the temperature to change from $y_1 = 350\,°C$ to $y_2 = 500\,°C$, that is, the process variable change was $\Delta y = 500 - 350 = 150\,°C$. This gives us the following calculation for the process gain:

$$K_p = \frac{\Delta y / PV_{range}}{\Delta u / OP_{range}} = \frac{150/1000}{10/100} = 1.5$$

□

Throughout this book, we will work with normalised signals. This means that when nothing else is stated, we assume that all signals are given as a percentage and that $OP_{range} = PV_{range} = 100\%$.

2.3 Common Process Types

The dynamic properties can be very different for various processes. The dynamics differ in speed, gain, stability, damping and more. In this section we will group different behaviours into certain common types. This grouping will make future discussions easier, as we here define some concepts and process types that we will use extensively later on.

It is important to specify what is included in the process as shown in the control loop in Figure 1.1. The process section consists of everything outside the controller, that is, all the dynamics that exist between the controller's output signal and the process variable. This means that, for example, sensors, transmitters and actuators are included in what we call "the process". In addition, filters and other functions, which are implemented in the controller or control system outside the controller module itself are included in the process.

Figure 2.2 shows the step response of six different process types. We will briefly discuss each of them in the following.

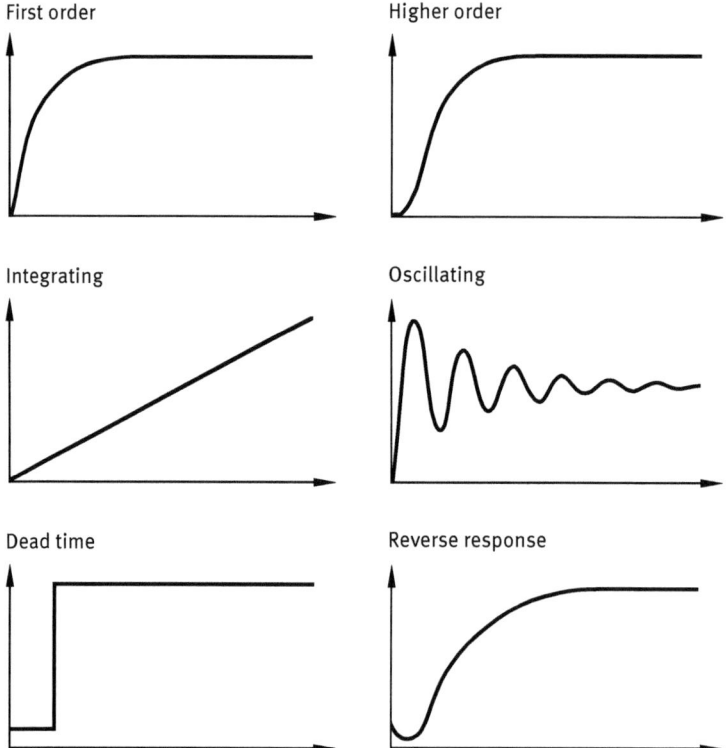

First order

Higher order

Integrating

Oscillating

Dead time

Reverse response

Fig. 2.2: Step responses of several process types describing the dynamic behaviour.

First Order Process

The first order process gets its name from the fact that it can be described by a first order differential equation. Its step response has the same appearance as the response of a simple RC circuit, that is, an electrical circuit consisting of a resistor and a capacitor when we switch on the sources. The time constant for an RC filter is $T = RC$ and the first order process is sometimes called a single-capacity process.

This is the simplest type of process we will encounter. Suppose we have a mercury thermometer with a very thin glass wall that is immersed in boiling water. The mercury column will then rise in the same way as the output of the RC filter and the relationship between the water temperature and the height of the mercury column can be described by a first order differential equation and hence a first order process. When passing a signal into a first order process it acts as a low-pass filter. This can be seen in the step response: The high frequency components of the sharp increase in the step are filtered out and the response has a smooth increase.

Higher Order Process

This is the most common type of process, often in combination with a dead time process as described below. The step response is the same as what you would get when you connect two or more first order processes in series. It is sometimes also called a multi-capacity process because it be described by a series of single-capacity processes.

Suppose we conduct a similar experiment with the same mercury thermometer as before, but with the difference that we now place the thermometer in a container with water at room temperature. We then place the filled container, with the thermometer inside, on a hot plate. The water temperature will now rise with a response corresponding to a first order process. However, the mercury in the thermometer is not affected by a step like in the previous example. Instead, the water temperature is the input signal that warms up the mercury in the thermometer. As a result, the mercury reacts more slowly. The thermometer shows a response which has the appearance corresponding to two first order processes.

The difference is that we now have two "capacities" to be heated. The difference between a first order and second order process becomes larger the more "capacity" it has. If, for example, the thermometer's glass wall is thick or if the hotplate is cold in the beginning, it will take longer to heat up the thermometer.

Integrating Process

Most processes in the process industry are stable. This means that if you change the controller output from one value to another, then the process variable will — after a while — adjust to the new level.

There are some processes that are of integrating type. These processes do not have the above property of stability. Examples are level control, pressure control in closed tanks, concentration control in batches and temperature control in well-insulated tanks. For example, opening the inlet valve to a tank, in which the inflow and outflow were previously balanced will cause the level to grow indefinitely without settling to any new level. Common to all these processes is that some form of storage takes place. In case of level, pressure and concentration control, this is storage of volume or mass. In case of temperature control, it is energy that is stored.

Stable and integrating process types have major differences in terms of step response analysis and control, which we will discuss later on in the book. It is therefore useful to have a feel for what types of processes are integrating.

Oscillating Process

This type of process is characterised by a step response that moves around its final stationary value. The process type is not very common among uncontrolled processes in the process industry. However, it is unfortunately common among *controlled* processes. One of the rare occurrences of oscillating process is concentration control of recycling liquids.

In mechanical systems, on the other hand, oscillating processes are common when using elastic materials, such as soft axles in a servomechanism or spring structures.

Dead Time Process

This process type characterises systems in which the process variable does not respond to the step change until a certain time — called the dead time — has passed. It is rare that the process consists of only a dead time, but the dynamics usually are combined with other process behaviours shown in Figure 2.2.

Dead times most often occur during material transports in pipes or conveyor belts. For example, we may want to measure the concentration of a liquid that is being transported in a pipe. This liquid is added some distance upstream in relation to the measurement. The dead time corresponds to the time it takes for the liquid to travel between the location of the addition and the location of the sensor.

Another process that is widely known is temperature control in a shower. A change in the mixture between cold water and hot water is only noticed when the water has been transported through the shower hose. Thus, we get a dead time which corresponds to the travel time of the water in the pipe.

Dead times also occur due to delays in modern instrumentation such as communication buses, networks and computers. Processes with long dead times are sometimes very difficult to control. Dead time processes are described in more detail in Section 6.8 which also covers some special methods for controlling them.

Process with Reverse Response

The last type of process in Figure 2.2 is not very common. It is characterised by a step response that initially goes in the "wrong" direction. It is easy to imagine that such a process can pose a huge problem for the controller.

A common process with reverse response in the process industry arises when controlling the level in steam boilers. If, for example, you want to raise the level of steam by increasing the water flow into the boiler, the first reaction is that the water in the vessel cools down. Steam bubbles in the water then disappear and, as a result, the level drops. Only after a while, when the water has warmed up again, does the level rise again to a higher level. Boiler level control is described in more detail in Section 6.5.

Another common process with reverse response are boilers fired with solid fuel. If you add more fuel, the first effect will be the cooling down of the fuel bed. The temperature will begin to rise only when the supplemented fuel starts burning.

Of course, the six process types mentioned above do not cover all processes that occur in industry. In particular, we have not taken into account any form of nonlinearity. We address this in Chapter 5. However, the vast majority of our processes can be roughly described using the above process types. Mostly, the processes are combinations of the six process types. This is illustrated in Figure 2.3 which shows step responses for processes that are combinations of two process types.

Of the processes described here, first order and integrating processes are the easiest to control. This is because the process variable immediately responds to a change in the controller output. For the same reason, the dead time process and the reverse response process are the processes that are the most difficult to control. This is because the process variable initially does not react at all or, worse still, reacts in the wrong direction after the controller output is changed. The oscillating process is easy to control if the specification do not ask for a fast response. If a fast response is required then one will get stronger oscillations.

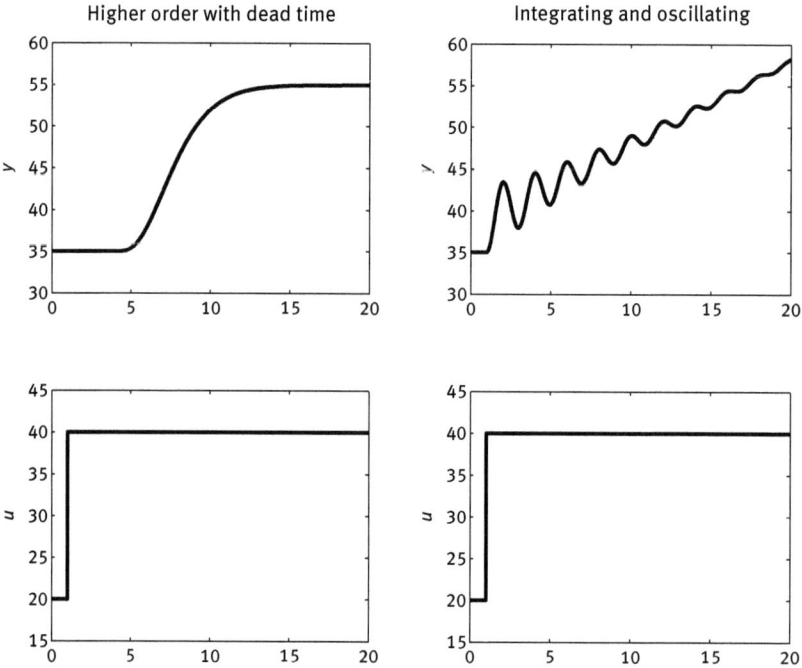

Fig. 2.3: Two step responses for processes that consist of two process types. The step response on the left comes from a higher order process combined with dead time and the response on the right from a combined integrating and oscillating process.

2.4 Step Response Analysis

Because the damping and the overshoot are normally zero, the other measures — dead time, time constant and process gain — are the most interesting features to describe the step responses as given in Section 2.2. Process descriptions based on the three the parameters K_p, L and T are by far the most common descriptions in the process industry and there are many tuning methods for PID-controllers that are based on exactly these parameters.

There are many ways to determine K_p, L and T. In modern controllers and control systems, there are methods that automatically determine these parameter, but mostly you have to analyse the step response yourself to get the parameters without any assistance.

Figure 2.4 shows the most common method to determine K_p, L and T, namely the so-called 63-percent method. Here, we consider a typical step response without any oscillation as shown in Figure 2.1. The parameters are determined as follows. First you find the point on the curve of the process variable where the slope is the steepest. You then draw a line through this point approximating the slope. Then you determine the

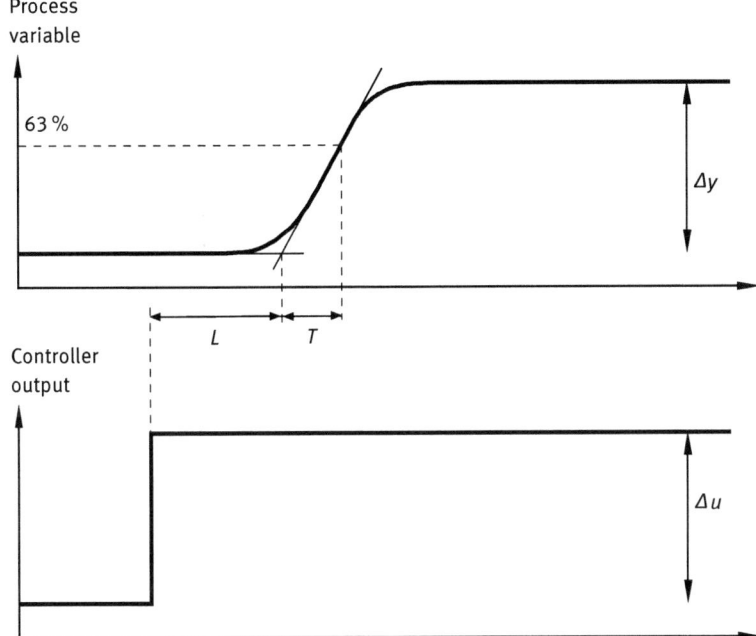

Fig. 2.4: 63-percent method for determining K_p, L, and T from the step response experiment.

intersection of this line with the constant value that the process variable had at the beginning of the experiment before the step change.

The time from when the step change was made to this intersection gives an estimate of the dead time L. From the intersection that marks L, the time constant T is then determined as the time it takes for the step response to reach 63 % of the final value. The static gain of the process K_p can finally be determined by dividing the change in the process variable y by the controller output change u:

$$K_p = \frac{\Delta y}{\Delta u}$$

We have previously defined the dead time as the time between the step change to the time the process variable begins to react. With the discussed method of determining the dead time, the *estimated* dead time L is usually longer than the *actual* dead time.

This is nevertheless a useful and correct estimation because of the very simple model of the process that we will use in the following. The dead time L and the time constant T are used to describe a process that actually consist of a dead time and several time constants. This is approximated with a slightly longer dead time and a dominant time constant.

The process gain K_p can be both positive or negative. For example, when controlling a temperature in a tank, K_p is positive if an increase in the controller output

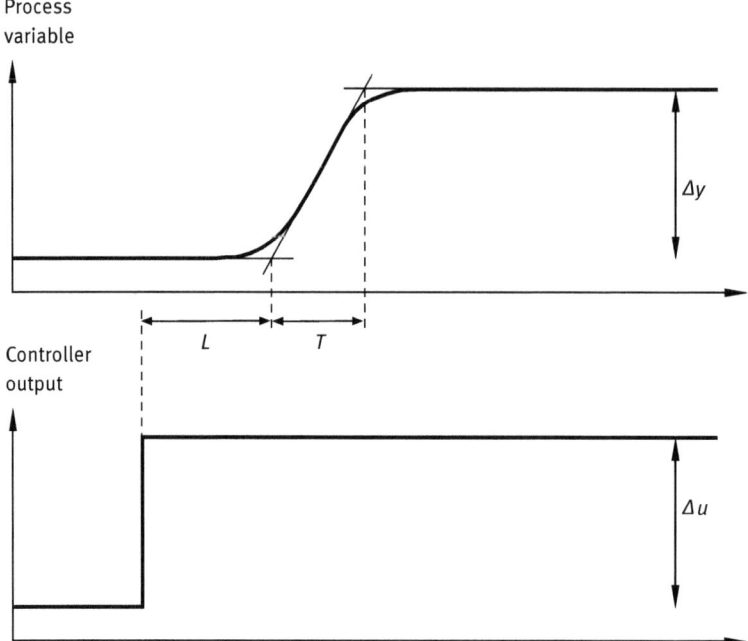

Fig. 2.5: Determining K_p, L, and T with a step response experiment.

increases the flow in a heating coil, but it becomes negative if the increase in the controller output increases the flow in a cooling coil.

A very similar method for determining K_p, L and T is shown in Figure 2.5. The only difference between the two methods is the way the dominant time constant T is determined. Instead of considering the time it takes to reach 63 % of the final value, one determines T as the time between the two intersections of the line drawn at the steepest slope with the stationary final value and the value of the process variable before the step change. This method is the basis for the Ziegler-Nichols tuning method described in Chapter 4. The method described in Figure 2.5 usually gives a time constant that is too long. In most contexts it is therefore recommended to employ the 63-percent method described in Figure 2.4.

Integrating Process

Integrating processes do not reach a new stationary level after a step change in the controller output. Figure 2.6 shows a step response experiment performed on an integrating process plus a first order process.

The first challenge is to ensure that the process is in steady-state, that is, finding a controller output, often a valve position, that gives a stationary process variable. When this is achieved, a step change is applied to the controller output. Because the

Process
variable

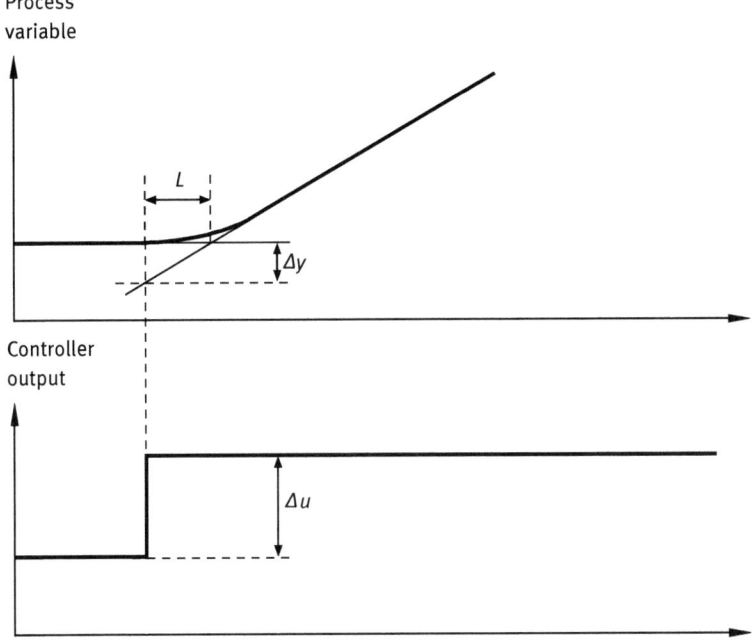

Controller
output

Fig. 2.6: Determining the process model of an integrating processes.

process is integrating, the process variable will not settle to a new level, but instead will, after a while, grow at a constant rate as shown in Figure 2.6.

An integrating process has an infinitely large steady-state gain K_p. For these processes, one therefore usually specifies the speed gain K_v instead, which tells how fast the process variable grows. From Figure 2.6, you can determine the speed of the process output as

$$v = \frac{\Delta y}{L}$$

The speed is of course proportional to the size of the controller output change Δu. The speed or velocity gain is therefore given as

$$K_v = \frac{v}{\Delta u} = \frac{\Delta y}{\Delta u \cdot L}$$

For stable processes we could determine two times, L and T, from the step response experiment. This is not possible for an integrating process. One usually makes do with determining the dead time L in the same way as before. That is, the dead time is determined by the intersection of the process value before the step change and the continuation of the slope caused by the integrating process. This is illustrated in Figure 2.6.

There are several tuning methods for PID-controllers based on the knowledge of K_v and L. This is described in more detail in Chapter 4.

Examples

We conclude this section by illustrating the methods for step response analysis on three simulated processes: a higher order damped process, an integrating process and a process with a long dead time.

Example 2.2. Step response analysis of a higher order damped process

Figure 2.7 shows the step response of a simulated higher order process. The simulated process consists of four filters, that is, four first order processes, in series where each filter has a time constant of one second.

Before the step response experiment is performed, the process is in steady-state. At time $t = 1$ s, the controller output is suddenly changed from $u = 20\%$ to $u = 40\%$. As a result, the process variable changes from $y = 35\%$ to $y = 55\%$. The static gain of the process can therefore calculated as

$$K_p = \frac{\Delta y}{\Delta u} = \frac{55 - 35}{40 - 20} = 1$$

Figure 2.7 shows the intersection between the steepest slope and the initial value of the process variable before the step change which is at $y = 35\%$. The intersection is at time $t = 2.43$ s. By measuring the time from when the step change was made we can determine the dead time as

$$L = 2.43 - 1 = 1.43 \text{ s}$$

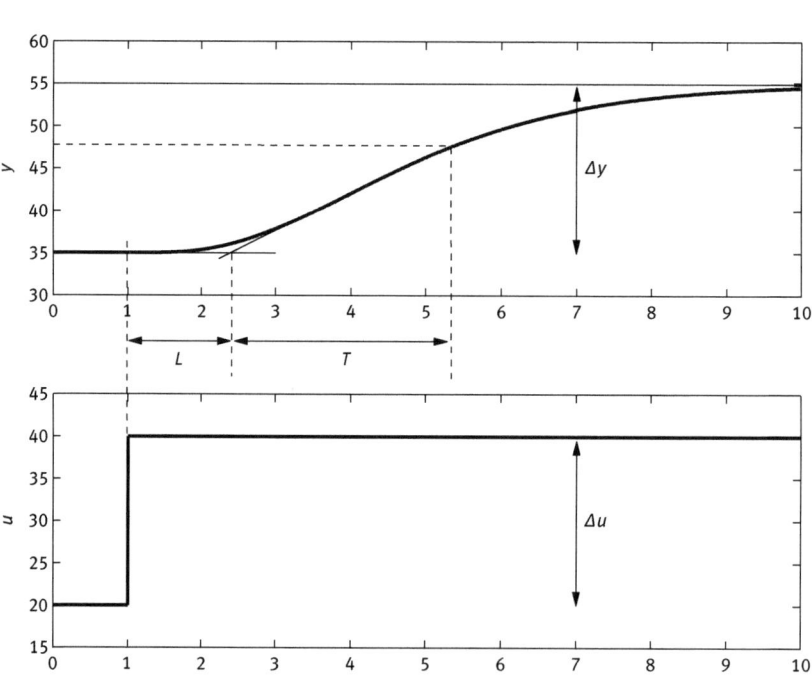

Fig. 2.7: Step response analysis of a higher order damped process from Example 2.2.

The process variable changes by Δy = 20 %. That means it has reached 63 % of its final value when it is at

$$y = 35 + 0.63 \cdot 20 = 47.6\%$$

In Figure 2.7 you can see that this happens at time t = 5.35 s. We can therefore determine the time constant as

$$T = 5.35 - 2.43 = 2.92 \text{ s}$$

If the method illustrated in Figure 2.5 for calculating the time constant had been chosen instead, the time constant would have been calculated as T = 4.5 s. ☐

Example 2.3. Step response analysis of an integrating process

Figure 2.8 shows a step response experiment performed on a simulated integrating process combined with two first order processes. The simulated process consists of two filters in series with a time constant of one second each, and an integrating process, also called integrator.

Before the step response experiment is performed, the process needs to be in steady-state. At time t = 1 s the controller output is suddenly changed from u = 20 % to u = 30 %. As a result, the process variable increases and after a while it grows at a constant rate. By drawing the slope in the process variable one can find the intersection of this line with the value of the process variable before the step change, y = 40 %.

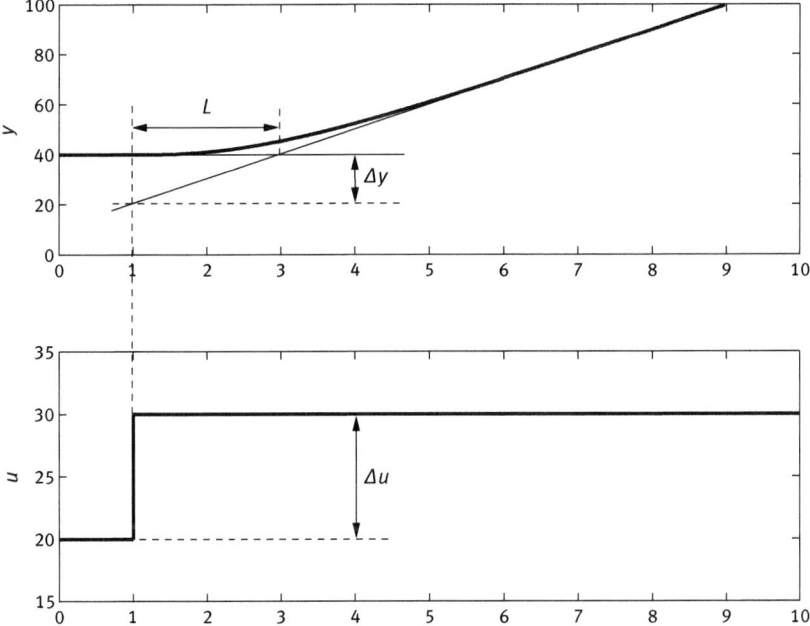

Fig. 2.8: Step response of an integrating process with two first order filters in Example 2.3.

The intersection occurs at $t = 3$ s. By measuring the time from when the step change was made to the time of the intersection one can determine the dead time as

$$L = 3 - 1 = 2\,\text{s}$$

Note that the process does not have an actual dead time. The combination of the integrating process with two first order processes constitutes the dead time of this model.

The controller output change is $\Delta u = 10\,\%$ and the change in the process variable can be determined as $\Delta y = 20\,\%$. From these measurements we can calculate the speed gain as

$$K_v = \frac{\Delta y}{\Delta u \cdot L} = \frac{20}{10 \cdot 2} = 1\,\text{s}^{-1}$$

$\square$

Example 2.4. Step response of a process with dead time

Figure 2.9 shows the step response experiment of a simulated process with a relatively long dead time. The simulated process consists of a higher order damped process described in Example 2.2 to which an additional dead time of a few seconds is added.

The process has to be in steady-state before the step response experiment is conducted. At time $t = 1$ s the controller output is suddenly changed from $u = 20\,\%$ to

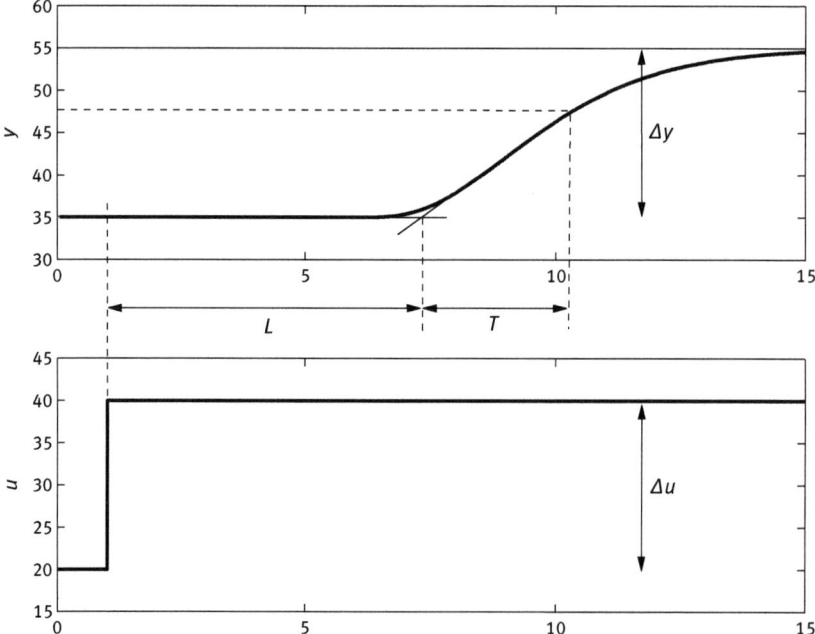

Fig. 2.9: Step response analysis of the dead time dominated process of Example 2.4.

$u = 40\,\%$. As a result, the process variable changes from $y = 35\,\%$ to $y = 55\,\%$. The static gain of the process can therefore be calculated as

$$K_p = \frac{\Delta y}{\Delta u} = \frac{55 - 35}{40 - 20} = 1$$

In Figure 2.9 the line corresponding to the slope of the process value is drawn. The intersection of this line and of the process variable level before the step change, $y = 35\,\%$, determines the time $t = 7.43$ s. By measuring the duration from when the step change was made to this time one can determine the dead time as

$$L = 7.43 - 1 = 6.43 \text{ s}$$

The process variable changes by $\Delta y = 20\,\%$. That means it has reached $63\,\%$ of its final value when it is at

$$y = 35 + 0.63 \cdot 20 = 47.6\,\%$$

In Figure 2.9 you can see that this happens at time $t = 10.35$ s. We can therefore determine the time constant as

$$T = 10.35 - 7.43 = 2.92 \text{ s}$$

If we compare these values with those obtained for the higher order damped process in Example 2.2 we see that the only difference is that the dead time has increased by five seconds, from 1.43 s to 6.43 s. □

2.5 Other Analysis Methods

A step-response experiment followed by a step-response analysis is the most prominent method for examining the dynamics of processes. It is simple and gives a good description of the dynamics. This description suffices for tuning PID-controllers in the process industry.

There are, of course, other methods to describe and determine process dynamics. A relatively simple method is to induce an oscillation in the control loop and to deduce the dynamics of the process from the oscillation frequency and amplitude. These methods are also useful when tuning PID-controllers and are described in more detail in Chapter 4.

Sometimes you want a more detailed description of the process dynamics. This applies, for example, when you want to build a process simulator or if you have more complicated controllers than the PID-controllers to be designed and tuned. There are many — both simple and advanced — methods to generate such models, but the description of this is outside the framework of this book. There is also good computer support for this type of model building. You will find many textbooks for those methods under the topic 'system identification'.

3 PID-Controllers

3.1 Introduction

We now move on to the second component in the single control loop as described in Figure 1.1 — the controller. There are many different types of controllers, but we will limit ourselves here to the PID-controller, which is the controller used most widely in the process industry.

There have been PID-controllers in various forms for several centuries, but it was not until the 1940s that the production of general purpose PID-controllers gained momentum. The design of the PID-controller has changed over the years. The first PID-controllers were pneumatic. Then came the analog electrical controllers and in the 1970s these evolved to computer-based digital controllers. At the same time, control systems that contained several PID-controllers were introduced. Together with many other functionalities, the PID-controllers form building blocks in the industrial control systems. Despite these major changes in manufacturing technology, the basics function of the controller, the PID algorithm, has not changed. It probably never will change either.

In this chapter we describe the structure of the PID-controller and how it functions. If you understand what each of the three terms in the controller — the P-, I- and D-part — does, it is much easier to use the entire controller. First of all, one can deduce when the different parts are needed, i.e. when to use a P-, PI-, PD- or PID-controller. Secondly, it is easier to tune the controller parameters if you understand the function of the individual parts. We discuss the tuning of the PID-controller in Chapter 4.

3.2 PID-Controller Structure

We first derive the structure of the PID-controller and show that it is a natural extension of the simplest form of control, namely the on/off controller.

On/off Controller

The on/off controller is the simplest form of controller one can imagine. Its output signal u is given by

$$u = \begin{cases} u_{max} & e > 0 \\ u_{min} & e < 0 \end{cases} \tag{3.1}$$

where e is the control error, that is, the difference between setpoint and process variable.

$$e = r - y$$

https://doi.org/10.1515/9783111104959-003

The on/off controller can also be described graphically as shown in Figure 3.1. Here, we assume that the process has a positive process gain, that is, an increase in the controller output causes an increase in the process variable. The controller output of the on/off controller can only assume two values, u_{max} and u_{min}. Depending on whether the control error is positive or negative, the high and low levels are selected as the controller output.

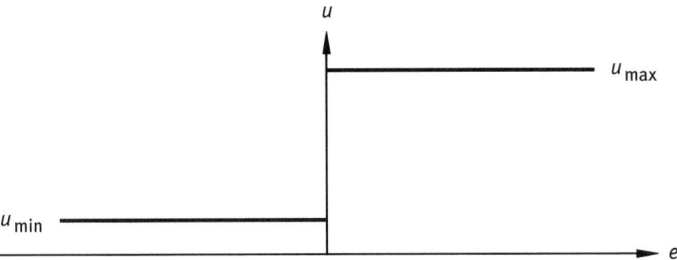

Fig. 3.1: Controller output of an on/off controller.

Example 3.1. On/off Controller

Figure 3.2 shows control with an on/off controller. The process is the higher order process described in Example 2.2. The setpoint is $r = 50\,\%$ and the controller output switches between the levels $u_{min} = 10\,\%$ and $u_{max} = 90\,\%$. The figure shows that the process variable is roughly at the desired setpoint, but is still oscillating around the setpoint. □

A major disadvantage of the on/off controller is that it causes oscillations in the control loop. This is illustrated in Figure 3.2. The controller must constantly switch the output signal between the two levels u_{max} and u_{min} to successfully keep the process variable close to the desired setpoint. If, for example, we control a level in a tank using a valve that can only assume the positions open and closed, we must of course alternately open and close the valve to keep the average level at the setpoint. These variations in the controller output causes oscillations in the process variable and often even in adjacent control loops that are disturbed by this unwanted control behaviour.

P-Part

The function of the on/off controller works well for large control errors. It then often makes sense to either open or close the valve completely. The oscillations are caused by the controller's behavior in the event of small control errors. To avoid oscillations, the controller gain should be small for small values of the control error. This can be

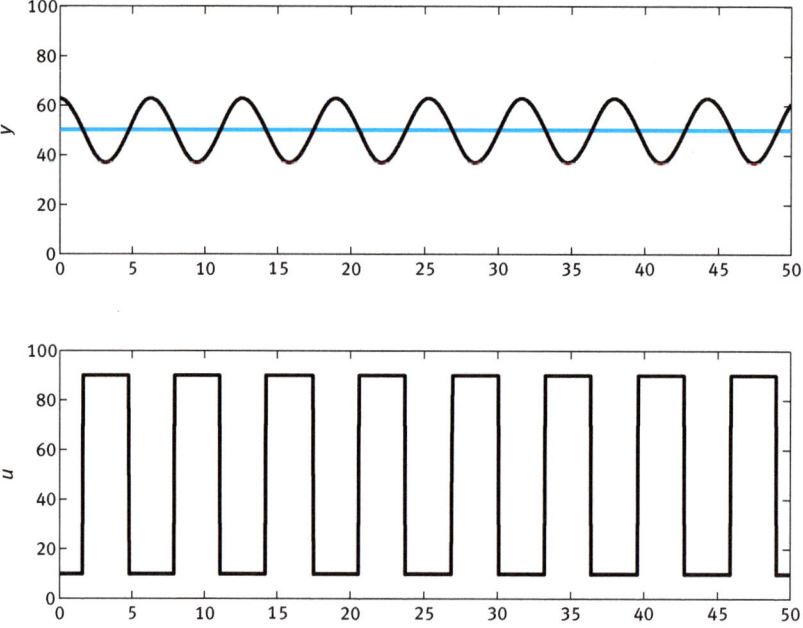

Fig. 3.2: On/off control of a higher order process. The upper graph shows the setpoint $r = 50\%$ and the process variable y. The lower diagram shows the controller output u. The limits of the controller output are $u_{min} = 10\%$ and $u_{max} = 90\%$.

done by introducing a proportional band, or a P-controller. The controller output of a P-controller is

$$u = \begin{cases} u_{max} & e > e_0 \\ u_0 + Ke & -e_0 \le e \le e_0 \\ u_{min} & e < -e_0 \end{cases}$$

where u_0 is the level of the controller output when we have no control error and K is the gain of the controller. The P-controller can also be described graphically according to Figure 3.3.

The output of the P-controller corresponds to the output of the on/off controller in case of very large control errors, that is, for absolute values of e greater than e_0. In all other cases the controller output becomes proportional to the control error.

Example 3.2. P-control

Figure 3.4 shows control with a P-controller of the higher order process described in Example 3.1. The figure shows control with two different controller gains, $K = 0.5$ and $K = 2$. The setpoint is constant $r = 50\%$. At time $t = 5$ the process is disturbed by a load disturbance, i.e. a disturbance which is added to the controller output. Load disturbances are described in more detail in Section 4.2. The figure shows that the

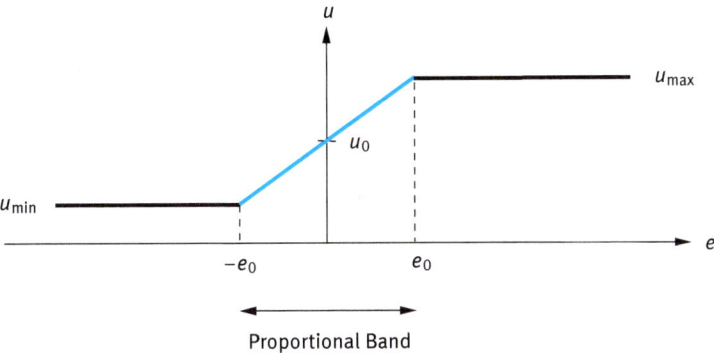

Proportional Band

Fig. 3.3: Controller output of a P-controller

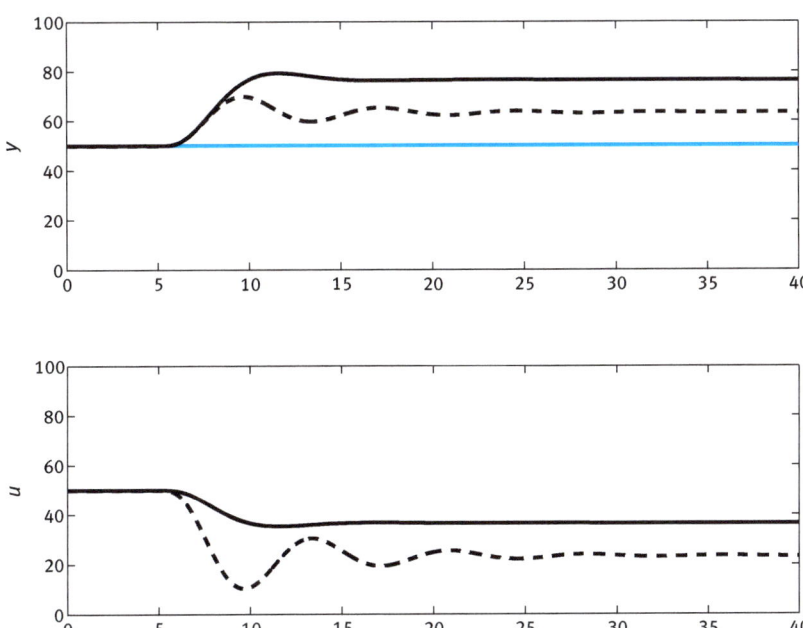

Fig. 3.4: P-control. A load disturbance occurs at $t = 5$. The upper graph shows the setpoint $r = 50\,\%$ and the process variable y. The lower diagram shows the controller output u. The figure shows two cases with two different gains, $K = 0.5$ (solid) and $K = 2$ (dashed).

persistent oscillation that existed for the on/off controller has disappeared. The figure also shows that the process variable does not return to the desired setpoint after the load disturbance has occurred. Instead, we see a remaining control error, also referred to as *offset*. The higher the controller gain K, the smaller the offset. On the other hand, the control becomes bumpier when the gain is high, exhibiting a more pronounced oscillation. □

The P-controller eliminates the oscillation of the on/off controller. Unfortunately, this happens at the price of getting a new problem: The steady-state control error does not necessarily become zero or, in other words, setpoint and process variable are not the same after all short-term dynamics have disappeared.

This can be explained by studying the expression of the controller output. In the event of small control errors, the P-controller works within its proportional band, that is, the relationship between the controller output and the control error is given by

$$u = u_0 + Ke$$

In other words, the control error is given by

$$e = \frac{u - u_0}{K}$$

In the steady-state case, the control error will only be zero ($e = 0$) if $u_0 = u$. So we can only get rid of the steady-state control error if we can adjust the value of u_0 so that u_0 assumes the same value as the controller output u for all values of the setpoint r.

Because the control error of the P-controller is inversely proportional to the controller gain K, we see that the higher the controller gain, the smaller the control error. We also see that we minimise the largest steady-state control error that can occur if we select a value of u_0 located in the middle of the controller output operating range. In most industrial controllers u_0 is therefore pre-selected to $u_0 = 50\,\%$. In some controllers you can adjust the value of u_0. From the discussion above, we see that in this case one should choose u_0 as close to the desired stationary value of the controller output u as possible. For example, a valve may be designed so that it is mostly open at 70 %.

I-Part

Instead of keeping u_0 constant, one can try to adjust it automatically so that it fulfills $u_0 = u$ after all variables in the control loop have settled down. The steady-state control error will be eliminated as a result. This is exactly what the integral part (I-part) of the PI-controller does. The controller output of a PI-controller is given as

$$u = K\left(\frac{1}{T_i} \int e(t)dt + e\right)$$

where T_i is the integral time of the controller. It is also sometimes referred to as the reset time. The constant value u_0 in the P-controller has thus been replaced by the term

$$u_0 = \frac{K}{T_i} \int e(t)dt$$

which is proportional to the *integral* of the control error. This is why the term is called the integral term or the integral part of the PID-controller. The integral of the control

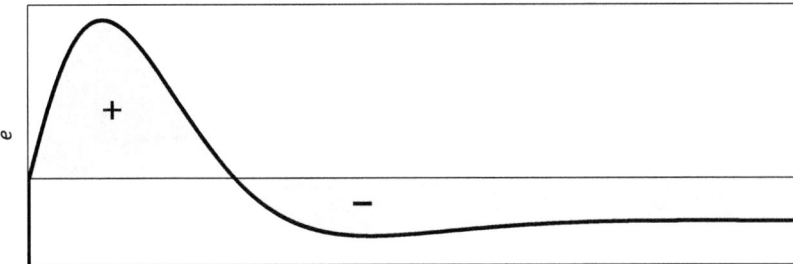

Fig. 3.5: The area under the control error curve is proportional to the integral part.

error is the same as the area under the curve of the process variable fluctuating around the setpoint. The integral part is proportional to this area under the curve as is illustrated in Figure 3.5. You can also approximate the integral part as the sum of all past control errors, weighted by the controller gain K divided by the integral time T_i, which we will discuss further in Section 3.6.

One can validate that the PI-controller has the ability to eliminate the remaining control error by studying the above control law. Assume that we control with a PI-controller and that we have a persistent control error when all signals have settled down, as shown in Figure 3.5. If the control error e is constant, the proportional part of the PI-controller will be a constant with value Ke. However, the integral part will not be constant. It will increase or decrease, depending on whether the control error is positive or negative, because the integral part constantly sums the control error e. This means that the controller output cannot be constant. As long as there is a control error, the integral part will increase or decrease. The only way to achieve a constant integral part and thus a constant controller output is when the control error becomes $e = 0$.

Example 3.3. PI-control

Figure 3.6 shows the control of the same higher order process as in Examples 3.1 and 3.2, this time controlled with a PI-controller. The controller parameters are calculated using the AMIGO method described in Chapter 4. They are $K = 0.414$ and $T_i = 2.67$ s. The figure also gives the results of a controller with P-control only with same gain, $K = 0.414$. The figure shows that the process variable with the PI-control returns to the setpoint after a load disturbance, i.e. there is no steady-state error in the process variable, while a large offset persists for the P-controller. □

We have now shown that a PI-controller solves both the problem with oscillations that occurred for the on/off controller and the problem with persistent control errors that occurred for the P-controller. The PI-controller is therefore a controller without any real shortcomings. It is usually sufficient when the requirements for the performance

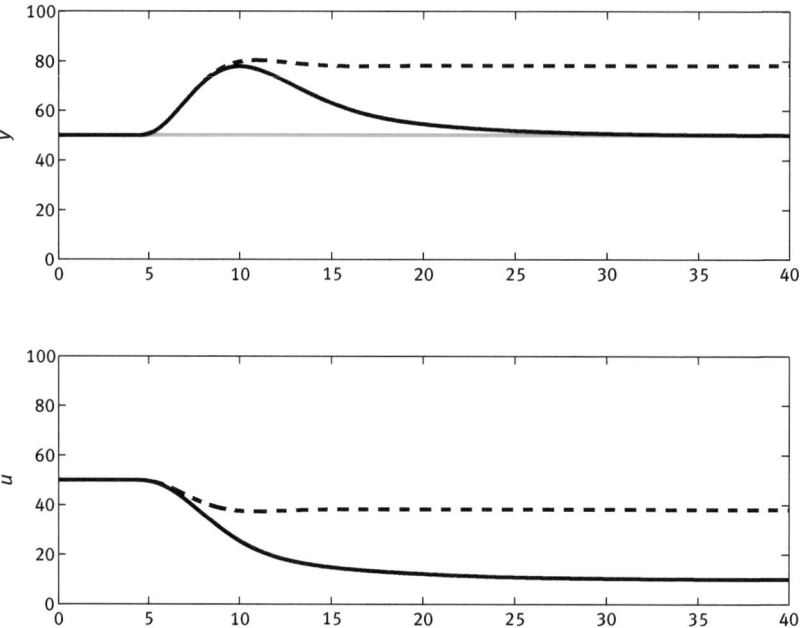

Fig. 3.6: PI-control. The control loop is disturbed by a load disturbance at time $t = 5$. The upper graph shows setpoint $r = 50\%$ and process variable y. The lower graph shows the controller output u. The dashed line shows control with a P-controller with the same gain as the PI-controller.

of the control loop are not too strict. That is why the PI-controller is so common in industrial applications.

D-Part

One feature that limits the performance of the PI-controller is that it only looks at past control errors and thus at what *has* happened. It does not try to predict what will happen to the control error in the near future. The problem is illustrated in Figure 3.7. Figure 3.7 shows the development of the control error for two different scenarios. The P-part of the controller is proportional to the control error at current time t. This control error is equal in both figures. The integral part is proportional to the area under the curve of the control error. These areas are also equal in both cases. This means that a PI-controller provides exactly the same controller output at time t in both cases. A good controller should see that there is a big difference between the two scenarios. In the plot on the left, the control error decreases rapidly and the controller should take careful action to avoid an overshoot in the control error. In the plot on the right, the control error has started to suddenly increase again after an earlier decrease. Here, the

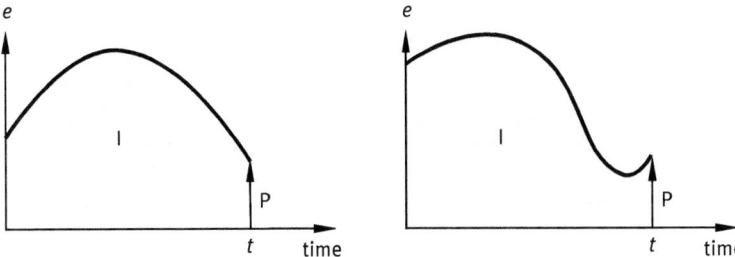

Fig. 3.7: Two control scenarios where the controller output from a PI-controller is similar at time t.

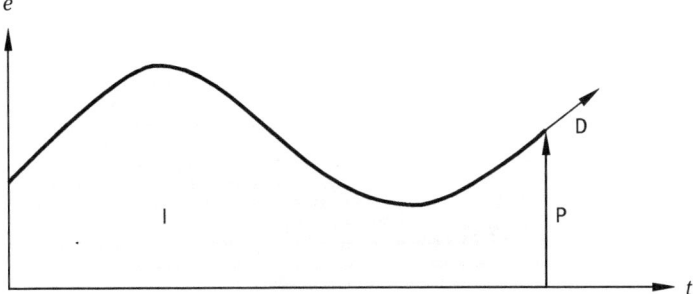

Fig. 3.8: The integral part is proportional to the area under the control error, the proportional part is proportional to the current control error and the derivative part is proportional to the change of the control error.

controller should take strong action so that the control error will subside again. The derivative part of the PID-controller performs exactly this type of compensation.

The D-part of a PID-controller is proportional to the change in the control error, that is, it is proportional to the derivative. This is illustrated in Figure 3.8.

The mathematical expression of the PID-controller is:

$$u = K\left(e + \frac{1}{T_i} \int e(t)dt + T_d \frac{de}{dt}\right)$$

where T_d is the derivative time of the controller.

Example 3.4. PID-control

Figure 3.9 shows the control of the same higher order process as in Examples 3.1 to 3.3 but now with a PID-controller. The controller parameters are derived using the AMIGO method described in Chapter 4. They are $K = 1.12$, $T_i = 2.41$ s, and $T_d = 0.623$ s. The figure also shows control with the PI-controller described in Example 3.3.

The figure shows that the PID-controller acts faster than the PI-controller when the load disturbance occurs and thereby reduces its effect on the process variable. ☐

As mentioned earlier, the derivative part is not necessary in most cases. You can usually manage with a PI-controller only. The greatest benefit of the D-part occurs in cases

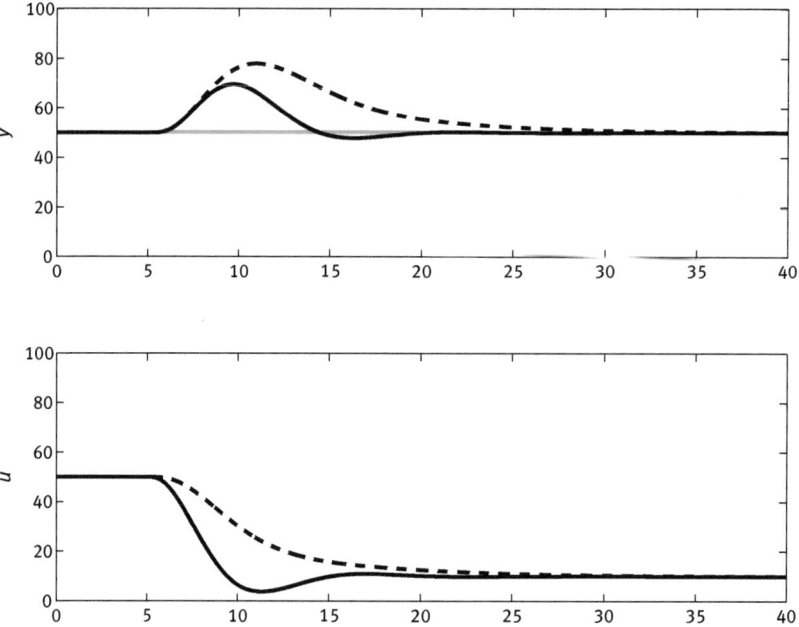

Fig. 3.9: PID-control. A load disturbance occurs at time $t = 5$ s. The upper graph shows the setpoint $r = 50\,\%$ and process variable y. The lower graph shows the controller output u. The dashed line shows control with the PI-controller as described in Example 3.3.

where you gain a lot from predicting the error. The signals must also be relatively free from noise, because the derivative part amplifies the noise even when using low-pass filtering. This is described in more detail in Section 3.6.

An example of cases where the derivative part is suitable is when controlling higher order processes, for example in many temperature control applications. Due to the sluggishness of the process, it is important to intervene well in advance in the heating process. Slow heat conduction can cause the temperature to rise long after the heating has stopped.

Anyone who has fried an egg using a thick-bottomed frying pan has seen this phenomenon. It can take quite a long time from the time you reduce the temperature of the stove to the time the temperature in the pan is lowered. Meanwhile, the temperature can overshoot significantly if you do not control it carefully enough. Sometimes, even DPID-controllers are employed, that is, controllers that not only have some part that is proportional to the first derivative, but also proportional to the second derivative of the process variable. This means that there is a part that is proportional to the *acceleration* of the process variable.

The PID-controller can be summarised as shown in Figure 3.8. The proportional part makes a contribution to the controller output that is proportional to the *present* control error. The integral part is proportional to the *past* and thus acts as the memory of the PID-controller. It is proportional to the sum of all past control errors. The derivative part tries to look ahead and is used to calculate how the control error will change in the near *future*.

Parallel and Series Form

Different controller manufacturers implement PID-controllers in different ways. This is mostly for historical reasons and has no major implication for the user as long as you tune the controllers manually by trial-and-error. However, with a systematic approach when tuning a PID-controller, it is important to know the structure of the controller. We will introduce the two most common realisations, namely the parallel form and series form.

Parallel Form

The parallel form was described earlier in the book. Here, the controller output is given by

$$u = K\left(e + \frac{1}{T_i}\int e(t)dt + T_d\frac{de}{dt}\right)$$

The parallel form is in some contexts also called the ideal form. It can be graphically described as shown in Figure 3.10.

The parallel form is the form normally described in textbooks. Characteristically, the P, I and D parts are separated. For historical reasons, this form was often not used in industrial process controllers because it was difficult to realise the parallel form with pneumatic elements in the old controllers. In many cases, people have not traditionally switched to this form even though it is as easy to implement as other forms in modern computer-based controllers. However, the parallel form is becoming more common in modern control system, probably because they did not have to consider historical or legacy systems.

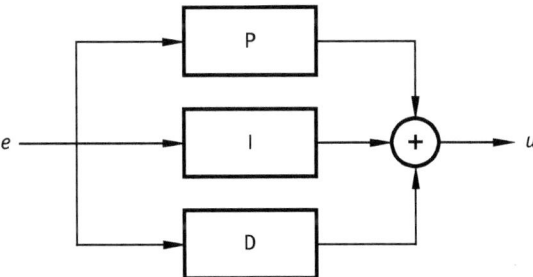

Fig. 3.10: Graphical description of the parallel form.

Series Form

The series form is a very common form in industrial process controllers. The controller output is given here by the following equations

$$e_1 = e + T'_d \frac{de}{dt}$$

$$u = K' \left(e_1 + \frac{1}{T'_i} \int e_1(t)dt \right)$$

Graphically, it can be described as shown in Figure 3.11. In the series form, the I- and D-parts are not independent as they were in the parallel form. The P- and I-part act both on the control error and the output signal from the derivative part. The controller can be seen as a series connection of a PI- and a PD-controller. In the English-language literature, the series and parallel forms are therefore often called *interacting* and *non-interacting* form, respectively.

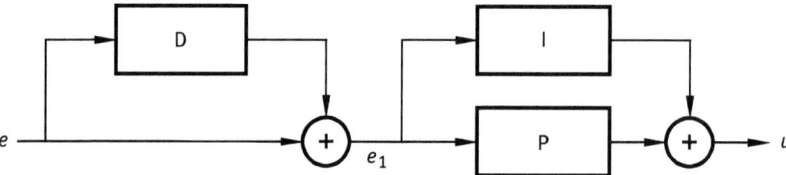

Fig. 3.11: Graphical description of the series form.

Relationship Between Parallel and Series Form

At first glance, the parallel form and the series form appear to be completely different control structures. However, it can be shown that this is not the case. They are both PID-controllers, that is, controllers with a proportional term, an integral term and a derivative term. The only difference is that the controller parameters K, T_i and T_d in parallel form do not have the same meaning as K', T'_i and T'_d in series form.

In the following, we describe the relationships between the PID parameters in the two realisations. If we have the parameters given in series form, we can always calculate the corresponding parameters in parallel form. They are given by:

$$K = K' \frac{T'_i + T'_d}{T'_i}$$

$$T_i = T'_i + T'_d$$

$$T_d = \frac{T'_i T'_d}{T'_i + T'_d}$$

If we have parameters given for a controller in parallel form, we cannot always calculate the parameters of the equivalent series form. The condition for it to work is that

$$T_i \geq 4T_d$$

This means that the parallel form is more general than the series form. On the other hand, the above relationship is often fulfilled and the relation between the parameters then becomes:

$$K' = \frac{K}{2}\left(1 + \sqrt{1 - \frac{4T_d}{T_i}}\right)$$

$$T_i' = \frac{T_i}{2}\left(1 + \sqrt{1 - \frac{4T_d}{T_i}}\right)$$

$$T_d' = \frac{T_i}{2}\left(1 - \sqrt{1 - \frac{4T_d}{T_i}}\right)$$

Note that the series form and the parallel form differ only when we have a PID-controller, that is, when all three parts of the controller are used. If we have a P-, PI- or PD-controller, the two forms are identical. You can also easily verify this yourself by studying the two Figures 3.10 and 3.11.

The different realizations are a problem when changing the controller manufacturer, make or version. One may then be forced to either redo the tuning or to recalculate the controller parameters according to the above formulas. Finally, it should be added that other factors can also cause a set of controller parameters that provide good control for one make to give poorer control in another make. Such factors are filtering and special logic for setpoint changes. This is described in more detail in Section 3.6.

3.3 Selection of Controller Types

After studying how the PID-controller is implemented and what functions the different parts have, we can now determine when one or more of its parts can be disabled.

On/off Controller

As we have seen, the simplest controller type, the on/off controller, has one great disadvantage: it causes oscillations in the process variable. It also has a controller output that can cause severe wear of the actuator in use, such as valves and pumps. The on/off controller also has great advantages. It is often cheap to manufacture and requires no adjustment of any controller parameters. Because the controller is only based on the sign of the control error and not on its size, the sensor can often be very simple. Thermostats are examples for such simple and inexpensive controllers.

The on/off controller is suitable for use in processes where oscillations are no major disadvantage, where cheap manufacturing is required and where you do not want to spend much time choosing the controller parameters. The result is that we see this form of controller in our simplest household appliances such as ovens, refrigerators, freezers, heaters and irons.

P-Controller

For certain types of processes, we can work with a high controller gain in the controller without any stability problems. Many first order and integrating processes are examples of that. We have seen previously that a high gain in a P-controller decreases the steady-state control error. We therefore do not need an integral part in these control cases if we can accept the steady-state error.

Another case when steady-state errors do not matter is the inner loop in cascade control. Here it is also often recommended to use a P-controller only. Cascade control is described in Section 6.4.

Integrating and first order processes directly respond to a change in the controller output. This is clearly seen for the step response of these two processes in Figure 2.2. It is therefore not necessary for these types of processes to predict the control error and to compensate in advance. The D-part is therefore not needed for these process types. Because a high controller gain also amplifies noise despite filtering, which is then present in the controller output, it is best not to use the derivative part for these processes. The D-part is also not needed if we do not have strict control performance specifications.

PD-Controller

A well-insulated thermal process has an almost integrating process behaviour. Any added energy is almost entirely used to raise the temperature in the reactor because energy losses are negligible. Even with these types of processes, we can work with large controller gains and therefore often do not need any integral part of the controller.

In a thermal process, energy is stored instead of mass as it was the case in level control. Unlike level control, thermal processes often contain further difficult dynamics in addition to the integrator. These difficult dynamics stem from the heat material transport. This means that it is often not enough to only use a P-controller, but one must supplement it with a derivative part. The derivative allows us to stop the energy supply in time.

The PD-controller is sensitive to noise, as it has a relatively high amplification at high frequencies. A reason that the PD-controller works so well in thermal processes is that we can often sense the process variable, i.e. the temperature, with a relatively low noise level.

PI-Controller

The PI-controller is by far the most common control structure in the process industry. When deriving the PID-controller, we saw that the PI-controller was the simplest form that did not entail any significant disadvantages such as oscillations or offsets. We can almost always manage without the D-part if we do not have strict requirements regarding to the control loop response time. Another case mentioned previously were first order and integrating processes. Here, the D-part does not give a major improvement in the control performance, but is most troublesome because it amplifies the noise.

Another case suitable for PI-control is when we have long dead times. Admittedly, this is a type of process where one reaps the biggest benefit from predicting future control errors. To try predicting how the process variable will change in the near future via the derivative of the process variable, however, is not a good method.

Due to the dead time, it takes time before a control intervention is visible in the process variable. For this type of process, it is therefore much better to predict the future of the process variable by studying the controller output in combination with a process model. This is called dead time compensation and is described in Section 6.8. If we do not have access to dead time compensation it is better to use a PI-controller than a PID-controller.

A third case where we should disable the D-part is when the process is disturbed by a high noise level. We should, of course, in the first place try to filter out the noise, but sometimes this is not enough. The D-part then gives a bad prediction and should be removed.

PID-Controller

The D-part often gives better control performance than a PI-controller. It gives us an opportunity to increase both the P and I part, i.e. to increase the controller gain K and decrease the integral time T_i. The D-part provides the greatest improvement in control when the process is a higher order process with not too much dead time. These processes are by far the most common in the process industry. However, the vast majority of processes are today controlled with PI-controllers. The main reason why the derivative part is not used is that you may not have the time or knowledge to tune this part.

However, the use of automatic tuning has increasingly led to higher order processes being controlled with a PID-controller. Automatic controller tuning is described
in Chapter 4.

3.4 PID-Controller Parameters

We have now derived and discussed the structure of the PID-controller. The mathematical expression for the parallel form of the PID-controller is

$$u = K\left(e + \frac{1}{T_i} \int e(t)dt + T_d \frac{de}{dt}\right)$$

The PID-controller has three main parameters that the user can determine, namely
the gain K, the integral time T_i, and the derivative time T_d. There are many other parameters that affect the performance of the PID-controller and some of these are discussed in Section 3.6. The controller parameters should be selected so that the controller provides good control of the process it is applied to. The next chapter describes
several methods for choosing the controller parameters. Here, we will briefly describe
the physical interpretation of the three controller parameters. The knowledge of this
can be very helpful when choosing the controller parameters.

Gain *K*

A P-controller that works within the proportional band has a controller output given
by

$$u(t) = u_0 + Ke(t)$$

If, a certain time, the control error changes with size Δe this will result in a change in
the controller output with size

$$\Delta u = K\Delta e$$

This is illustrated in Figure 3.12.

In other words, the gain K of the P-controller describes how much the controller
output should react to a change in the control error. Because the gain acts as a scaling
factor before all three parts of the PID-controller, the corresponding interpretation also
applies to the PID-controller.

If the process has a small process gain, we want the control error to result in a relatively large change of the controller output. This is because a change in the controller
output will lead to only a small change in the process variable. If, on the other hand,
the process gain is high, the controller output must react to changes in the controller
error very carefully. From this we can conclude that the controller gain K should be
chosen *inversely* proportional to the process gain K_p.

Controller output

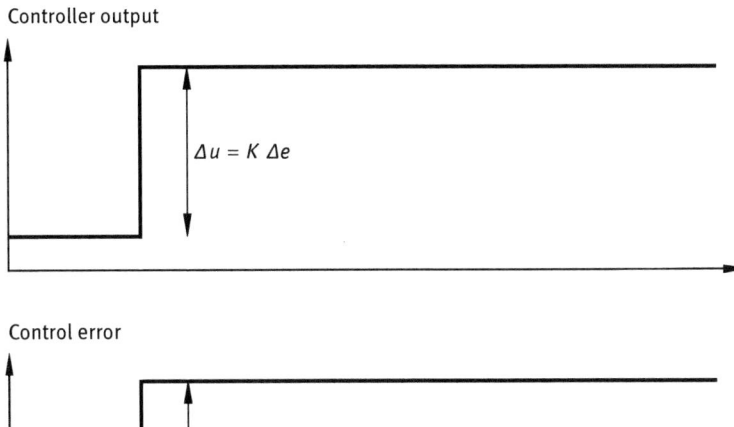

Control error

Fig. 3.12: Interpretation of controller gain K.

A positive control error will not necessarily cause a positive change in the controller output. If the process has a negative gain, that is, if $K_p < 0$ then an increase of the control error leads to a decrease of the controller output. This could be included by allowing the controller gain K to be negative, but this is almost never done. Most controllers and control systems assume a positive controller gain $K > 0$.

So instead, a logical variable captures the sign. If $K_p > 0$ you usually specify a *reversing* controller function and if $K_p < 0$ you usually specify a *direct* controller function.

In many controllers, the proportional band (PB), which was defined in Figure 3.3, is specified instead of the controller gain. The relationship between controller gain and proportional band is defined as

$$PB = \frac{100}{K} \, [\%]$$

For example, a controller gain of $K = 1$ is equivalent to the proportional band of $PB = 100\,\%$.

Integral Time T_i

The PI-controller has a controller output given by

$$u = K\left(e + \frac{1}{T_i} \int e(t)dt\right)$$

Controller output

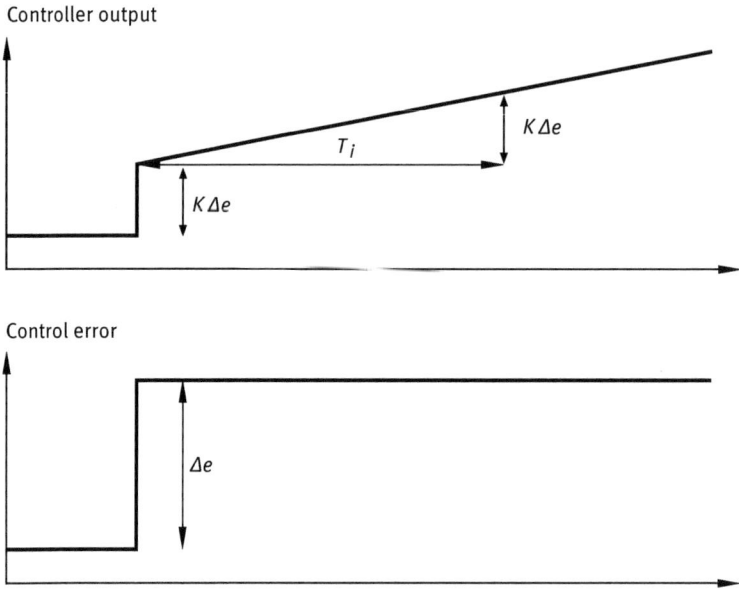

Control error

Fig. 3.13: Illustrating integral time T_i.

If the control error suddenly changes from zero to Δe the controller output changes by

$$\Delta u(t) = K\left(1 + \frac{t}{T_i}\right)\Delta e$$

where t is the time since the change in the control error. This is illustrated in Figure 3.13.

Just as with the P-controller, the controller output first changes with with a step of magnitude $K\Delta e$. After that, the controller output grows linearly with the speed determined by the integral time T_i. After time $t = T_i$, the controller output has grown to the value $K\Delta e + K\Delta e = 2K\Delta e$, that is, the step increase of the P-controller has doubled.

In other words, the integral time can be interpreted as the time it takes for the controller to double the step increase of the P-controller when the control error is constant. If the process is fast, we want this doubling to be done quickly, while a slow process requires the controller output to change more slowly. In other words, the integral time should be chosen proportionally to the dead time or time constant of the process.

The integral time has the dimension *time* and the unit is usually seconds or minutes. However, there are controllers and systems that instead use the unit rps which stands for repeats per second, or rpm which stands for repeats per minute. What this means is how many times the increase caused by the P-controller is repeated each second and minute, respectively. For example, if a controller has an integral time of

15 seconds, its unit can be specified as

$$T_i = 15\,\text{s} = \frac{15}{60}\,\text{min} = \frac{1}{4}\,\text{min}$$

or

$$\frac{1}{T_i} = \frac{1}{15}\,\text{rps} = \frac{60}{15}\,\text{rpm} = 4\,\text{rpm}$$

Derivative Time T_d

A PD-controller has a controller output that is given as

$$u = K\left(e + T_d\frac{de}{dt}\right)$$

The meaning of the derivative time T_d is described in Figure 3.14. If the control error — after a certain time — continues to increase or decrease in the same direction as its derivative, the following will apply:

$$e(t) + T_d\frac{de(t)}{dt} = e(t + T_d)$$

In this case, one can describe the controller output of the PD-controller as

$$u(t) = Ke(t + T_d)$$

Compared to a P-controller, where the controller output is proportional to the current value of the control error, the PD-controller gives a controller output that is proportional to what we think the control error will be at future time T_d.

In other words, the derivative part gives a prediction of the control error at time T_d in the future. For this prediction to be good, the change in the control error must be close to the straight line shown in Figure 3.14. It will do so if T_d is not too large. In fast

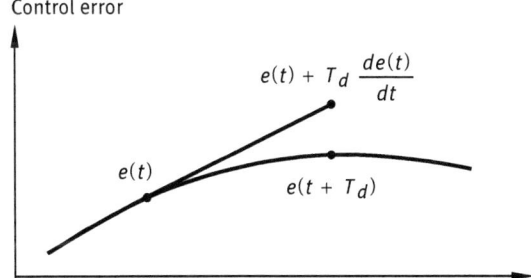

Fig. 3.14: Illustration of derivative time T_d.

processes, the control error changes quickly. In these cases T_d must be small. However, T_d can be chosen longer if the process is slow. In other words, the derivative time T_d should be chosen proportional to the dead time or time constant of the process.

The derivative time has the dimension time and the unit is usually seconds or minutes.

3.5 Identification of PID-Controllers

Identifying a PID-controller can seem strange at first. The controller is something you know and which you intentionally connect in your control loop to achieve good control. Unfortunately, the function of the controller is not always known.

Modern computer-based controllers are accurate by nature. The controller parameters that we set will indeed be the parameters used to control the process. However, the controller manufacturer does not always indicate if the controller structure is in series or parallel form.

For pneumatic and analog electrical controllers the situation is worse. We can not be sure that the control parameters indicated are those which are really used in practice due to aging and wear of the components.

Not knowing the exact function of the controller is not too problematic when you tune the controller manually by trial-and-error. If, on the other hand, you tune the controller using systematic methods, such as those described in Chapter 4, this uncertainty can result in drastically inferior control performance.

In this section, we describe how to use relatively simple means to identify the actual controller parameters and thereby be able to calibrate their scaling. A method to decide whether the controller is implemented in serial or parallel form is also presented, as well as a method for determining the process variable range PV_{range}. Common to all methods is that we experimentally study how the controller output reacts to variations in the process variable or in the setpoint.

Identification of Controller Gain K

The easiest way to determine the gain of the PID-controller is to disconnect the integral and derivative parts and to use the controller as a P-controller only. With some controllers you can disconnect the integral part completely. If this is not possible, you can make it small using a large T_i. The derivative part is removed by setting $T_d = 0$.

Figure 3.15 shows how to experimentally determine controller gain K. Insert a step change in the process variable and denote the size of the step Δe. Because the controller now acts as a proportional controller, the controller output also becomes a step, which size Δu is determined by the gain K:

$$K = \frac{\Delta u}{\Delta e}$$

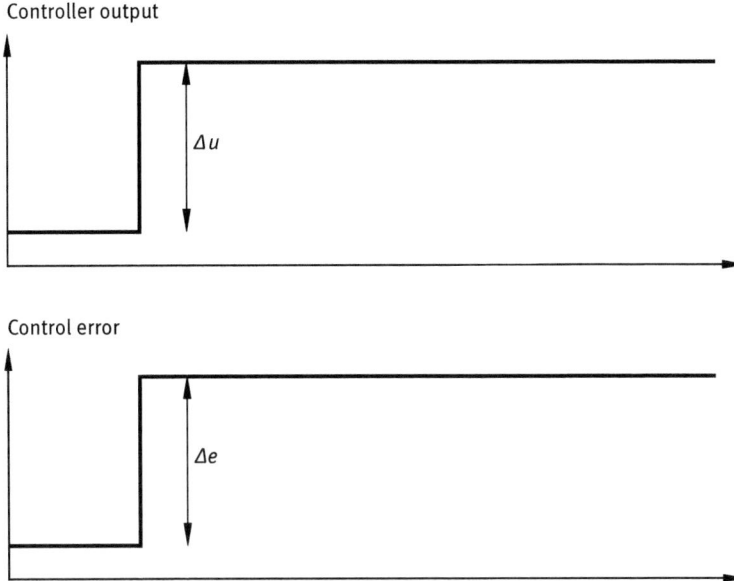

Controller output

Δu

Control error

Δe

Fig. 3.15: Experiment to identify controller gain K.

It is also possible to introduce a step change in the setpoint instead of the process variable, provided that the setpoint weighting is $b = 1$. Setpoint weighting is described in Section 3.6.

Identification of Integral Time T_i

When determining the integral time of the PID-controller, it is best to use the controller as a PI-controller and disconnect the derivative part by setting $T_d = 0$.

The controller output of a PI-controller increases (or decreases) as long as the control error is non-zero. The first step when identifying the integral time is therefore to stabilise the setpoint and the process variable so that the controller output becomes stationary.

Figure 3.16 shows how the integral time T_i can be identified experimentally. Starting in steady-state, a step change in the process variable will drive the control variable away. You can determine the integral time by studying how fast this effect is. With the notation indicated in the figure it is given as:

$$T_i = K\frac{\Delta e}{\Delta u}\Delta t$$

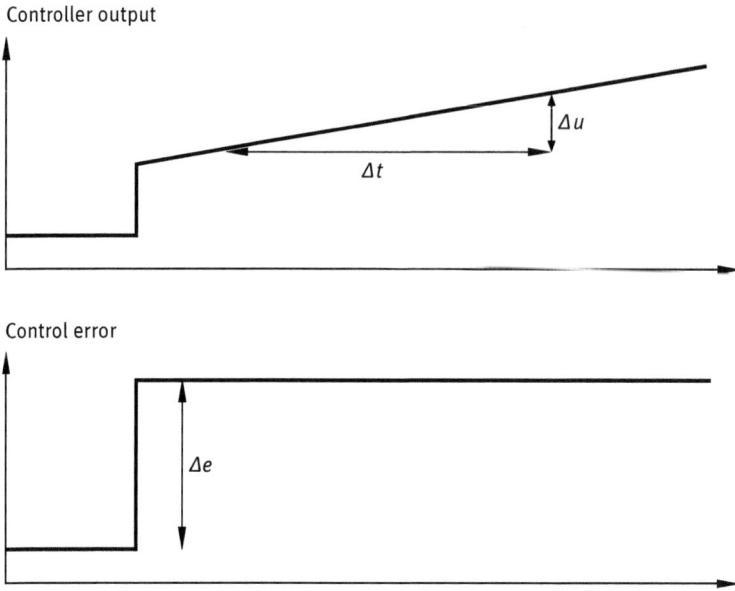

Controller output

Control error

Fig. 3.16: Experiment to identify integral time T_i.

Identification of Derivative Time T_d

The PID-controller's derivative time is easiest to determine when the controller acts as a PD-controller. It means the integral part must be disconnected or made very small by selecting a large value of T_i.

We were able to determine both the controller gain and the integral time through a step response experiment. The derivative time is difficult to determine from such an experiment because the derivative of a step is a short and high pulse. Instead, you have to determine the derivative time from a ramp response. This is shown in Figure 3.17.

If the process variable is a ramp, the controller output of the PD-controller will also become a ramp. The difference between a P-controller and a PD-controller is that the controller output of the PD-controller is proportional to the calculated error at future time T_d compared to the present control error. From the illustration shown in Figure 3.17 you can see how far ahead in the future this prediction is. Note that here a PID-controller with an additional first order filter is used. The filter removes the sharp increase that would otherwise occur when the control error starts to ramp up.

Identification of Controller Structure

The next chapter deals with different methods for tuning PID-controller parameters. There can obviously be large differences between the controller parameters depending

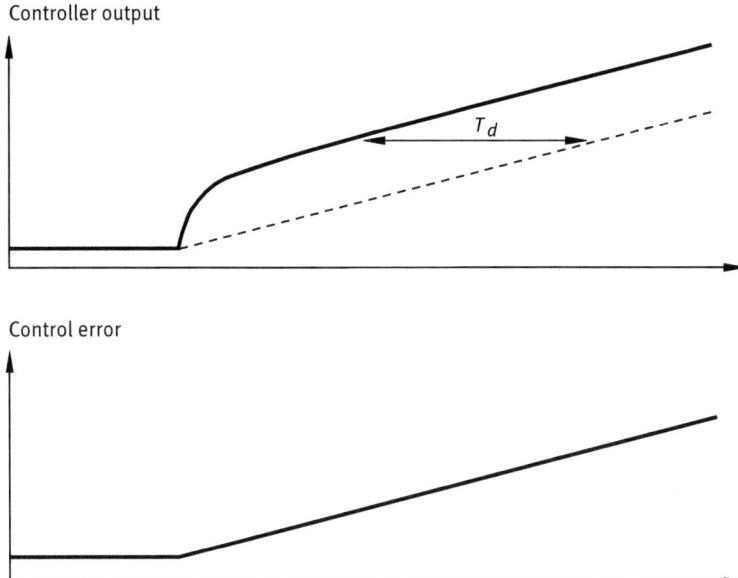

Fig. 3.17: Experiment to identify the derivative time T_d.

on which controller structure is used. It is therefore important to be able to determine the controller structure experimentally in cases where the structure is unknown.

The series form and the parallel form of the PID-controller are identical as long as not all three parts are active. The identification results of the previous sections are valid regardless of whether the controllers are implemented in series or parallel form because the experiments were made with P-, PI- and PD-controllers.

We therefore have to use a PID-controller to identify the controller structure. Similar to identifying the integral time, we have to start the identification by adjusting the setpoint and the process variable so that the controller output is stationary. We can then determine the structure by making a step change in the process variable and studying the response in the controller output as illustrated in Figure 3.18.

The distance Δu in the figure will be different depending on which controller structure is present.

For the parallel form, the following relationship holds:

$$\frac{\Delta u}{\Delta e} = K$$

For the series form; the following relationship holds instead:

$$\frac{\Delta u}{\Delta e} = K\left(1 + \frac{T_d}{T_i}\right)$$

For example, if you select $K = T_i = T_d = 1$, the ratios between the controller output change and the process variable change is 1 for the parallel form and 2 for the series form.

Controller output

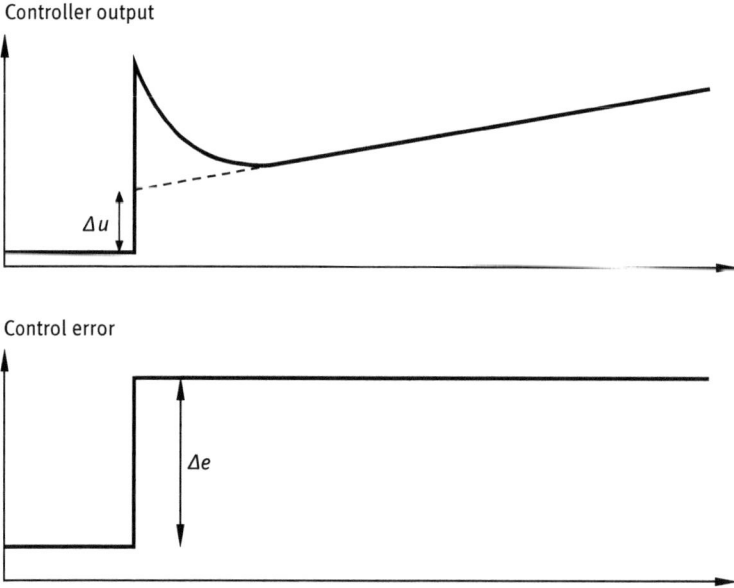

Control error

Fig. 3.18: Experiment to identify the controller structure.

Identification of Process Variable Range PV_{range}

In the previous chapter, we saw that you need to know the process variable range PV_{range} to determine the process gain K_p. Mostly, it is quite easy to determine PV_{range} in the control system, but in cases where this is difficult, you can determine the range experimentally. The method is illustrated in Figure 3.19.

Start by selecting the structure of the controller to a P-controller only with the appropriate controller gain K and wait until all variables are stationary. Then you need to introduce a step change in the control error of size Δe. The simplest way to achieve this is if you can do a step change in the setpoint r, but this assumes that the setpoint weighting is $b = 1$. If the controller is a real P-controller, b will be $b = 1$. However, sometimes the integral part cannot be disabled and you end up with a PI-controller with a very long integral time instead. If this happens, there is a risk that $b \neq 1$. In this case, a step change has to be introduced in the process variable.

Since we have a P-controller, a step change in the control error will cause an initial step change in the controller output with magnitude Δu. Since the controller works with normalised signals, this controller output change is given by

$$\frac{\Delta u}{OP_{range}} = K \frac{\Delta e}{PV_{range}}$$

This means that the process variable range can now be determined as

$$PV_{range} = K \frac{\Delta e}{\Delta u} OP_{range}$$

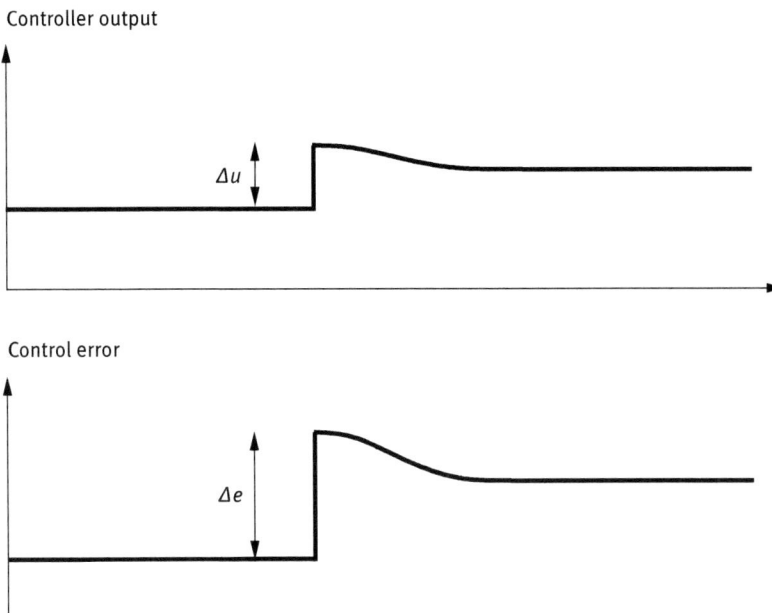

Controller output

Δu

Control error

Δe

Fig. 3.19: Experiment to identify the process variable range PV_{range}.

Assume that we have, for example, a temperature control loop with an unknown process variable range. The controller is set to P-control with controller gain $K = 0.5$. Then a step change is introduced in the setpoint with magnitude $\Delta r = \Delta e = 100°$ C, which results in a step in the controller output with the magnitude $\Delta u = 10\%$, where OP_{range} is 100 %. From the experiment we can determine the process variable range as

$$PV_{\text{range}} = K\frac{\Delta e}{\Delta u}OP_{\text{range}} = 0.5\frac{100}{10}100 = 500°C$$

3.6 Practical Modifications of PID-Controllers

The equation for the parallel form of the PID-controller is given by

$$u = K\left(e + \frac{1}{T_i}\int e(t)dt + T_d\frac{de}{dt}\right)$$

To get a practically useful controller this form should be modified in a few different ways. In this section, we will go through the most important modifications.

Setpoint Weighting

In the equation above, the controller output is determined by the control error e. The control error is given by $e = r - y$, i.e. the difference between the setpoint and the process variable. In practice, you usually have access to both these signals.

In many contexts, the setpoint is a signal that is constant for long periods of time and between these periods it may change suddenly. These instantaneous changes mean that the control error also changes suddenly. The derivative part of the PID-controller will react strongly to these changes and in most cases the controller output reaches its physical limit. You want to avoid this. Thus, you do not take the derivative of the control error but instead the derivative of the process variable. The structure of the PID-controller will then be

$$u = K \left(e + \frac{1}{T_i} \int e(t)dt - T_d \frac{dy}{dt} \right)$$

This controller structure still leads to sudden changes in the controller output when the setpoint changes abruptly. This is caused by the proportional part and is especially severe when the controller gain K is large. For this reason it is often necessary to change the structure also for the P-part so that

$$u = K \left(br - y + \frac{1}{T_i} \int e(t)dt - T_d \frac{dy}{dt} \right)$$

where b is called the setpoint weight and is a number in the interval $0 \le b \le 1$. In many controllers, b can only assume the values 0 or 1, but it is increasingly common to be able to select b over the entire interval between 0 and 1.

Avoiding sudden changes in the controller output is not the only reason why one would like a setpoint weight less than $b = 1$. Controllers that provide good control for load disturbances often give overshoots for abrupt setpoint changes. By making the setpoint weight smaller, you can reduce these overshoots.

Example 3.5. Setpoint weighting
Figure 3.20 shows the control of the same process with the same tuning as the PID-controller described in Example 3.4. Here, we study a setpoint change instead of the reaction to a load disturbance. The setpoint is not included in the derivative part. In the proportional part, setpoint weighting is used and the figure shows three cases: $b = 0$, $b = 0.5$ and $b = 1$. To completely avoid sudden changes in the controller output, the setpoint weight needs to be set to $b = 0$. This is a common choice in industrial controllers. The figure also shows how the overshoot decreases as the setpoint weight decreases. □

Setpoint weighting has a particular purpose when controlling an integrating process, namely to eliminate overshoots that occur when controlling for setpoint changes. It can be shown that it is not possible to avoid overshoots in case of setpoint changes

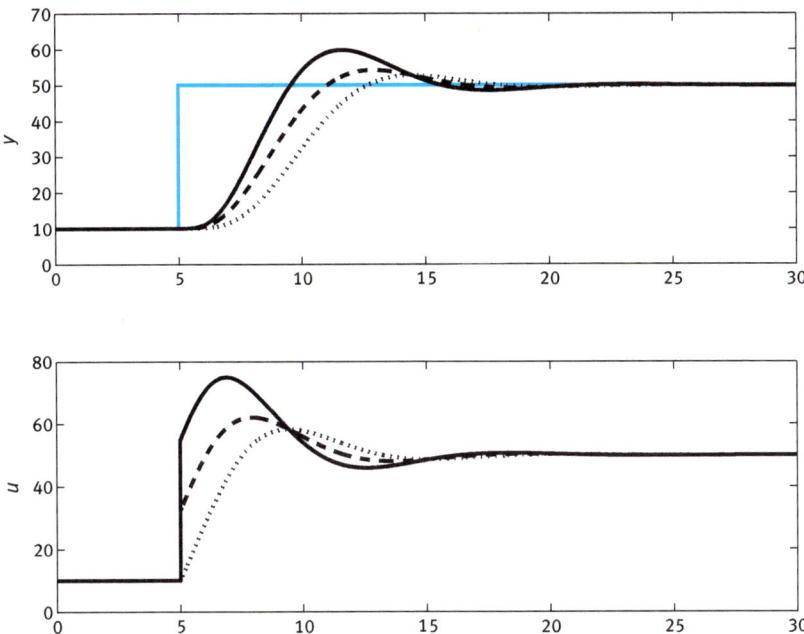

Fig. 3.20: PID-controller with a step change in the setpoint at $t = 5$ in Example 3.5. The figure shows closed loop control with three values of the setpoint weight: $b = 0$ (dotted), $b = 0.5$ (dashed) and $b = 1$ (solid).

when the process is integrating and the setpoint weight is $b = 1$. This is illustrated in the following example.

Example 3.6. Setpoint weighting for an integrating process

Figure 3.21 shows control of the same process as in Example 2.3. The controller is a PI-controller with integral time $T_i = 25$ s. The figure shows closed loop control with two different values for the controller gain, $K = 0.15$ and $K = 0.35$.

The figures on the left show control with a setpoint weight of $b = 1$. If a process is integrating and the setpoint weight is $b = 1$ the following can be shown: The area under the curve between the setpoint and the process variable when the process variable is *below* the setpoint is equal to the area when the process variable is *above* the setpoint. This can be seen in the figure and is why it is not possible to avoid overshoots when $b = 1$.

If you choose a setpoint weight that is less than $b = 1$, however, you can avoid overshoots for an integrating process. This is shown in the figures on the right-hand side where the setpoint weight is lowered to $b = 0.7$. □

The choice of setpoint weighting only matters for setpoint changes but does not affect the control performance when compensating for load disturbances. It should also be

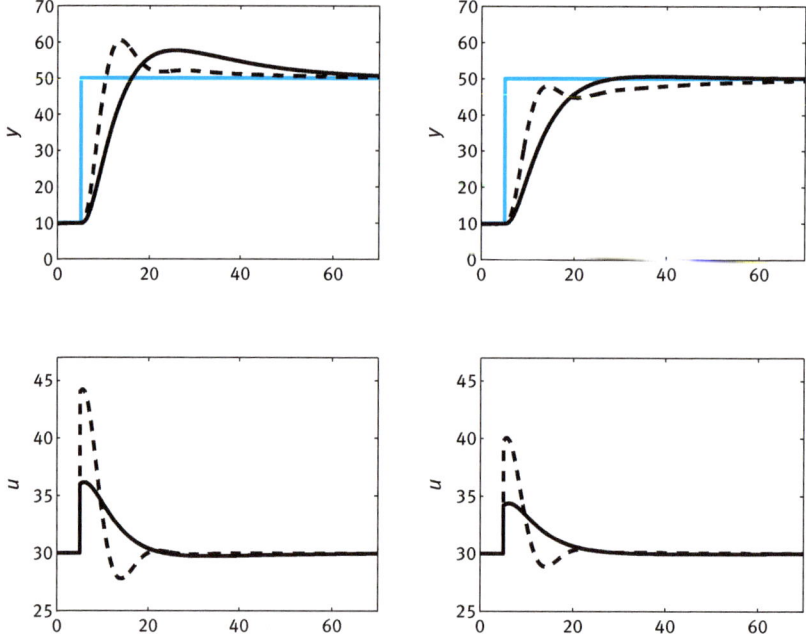

Fig. 3.21: PI-controller with a setpoint step change at time $t = 5$ for the integrating process of Example 3.6. The figure shows control with two different values of the controller gain: $K = 0.15$ (solid) and $K = 0.35$ (dashed). The figures on the left shows control with a setpoint weight of $b = 1$ and the figures on the right with a setpoint weight of $b = 0.7$.

pointed out that sometimes you are not content with setting $b = 0$. To get an even smoother reaction to setpoint changes, you can either let the setpoint pass through a low-pass filter or a ramp function before the signal is processed by the PID function.

Filters

In the previous section, we modified the PID-controller so that we got a good behavior in the event of abrupt setpoint changes. We must also be able to handle sudden changes in the process variable. Sudden changes in the process variable are primarily caused by measurement noise, but they can also be caused by quantisation. When we work with digital signals, there is always a certain resolution in the signal, that is, the signal jumps between different distinct levels. Normally, the resolution is so high that you do not see any quantisation effects. However, the quantisation becomes noticeable when you have poor resolution in the analogue-to-digital converter or if you have chosen a too large process variable range for the process variable. Quantisation is described in more detail in Section 6.2.

Quantisation effects should be eliminated by increasing the resolution of the analogue-to-digital converter or reducing the process variable range. If the quantisation effects cannot be eliminated, they must be regarded as measurement noise and handled in the same way as other measurement noise. Measurement noise is an unwanted component in the process variable that should be filtered out before the signal reaches the PID function. In modern controllers and control systems, low-pass filters can be applied to the signals. You should choose a filter time constant that is long enough to filter out measurement noise, but at the same time not so long that you make the process unnecessarily slow. Filtering of signals is described in Section 6.2.

The derivative part gives a signal that is proportional to how fast the process variable changes and is thus particularly sensitive to measurement noise. It means that high-frequency noise is strongly amplified by the D-part. In order to prevent the amplification of noise, a special low-pass filter is often applied to the derivative part of the controller.

The low-pass filter must ensure that the derivative part only acts on signals within the interesting frequency range. What is the interesting frequency range varies from case to case. Therefore, the low-pass filter must have different filter time constants that depend on how fast the process is. The most common method to solve this problem is to relate the time constant of the filter to the derivative time:

$$T_{\text{filter}} = \frac{T_d}{N}$$

In most industrial control systems N is fixed with N in the range of 5–10. In cases where the user must enter a value of the filter time constant the resulting filter time constant T_{filter} using this range of N and the equation above is therefore often an appropriate choice. In some cases, you are asked to enter a maximum gain at high frequencies and not the filter time constant. This maximum gain corresponds to the number N.

The filtering found in today's industrial PID-controllers is often not sufficient. Firstly, one often filters only the derivative term, and secondly, this filter is often not effective enough. If there are no filters acting on the proportional part, the measurement noise will be transmitted to the controller output amplified by a factor K, where K is the controller gain. It is therefore useful to low-pass filter the process variable before it reaches the PID function. If you have a PID-controller, the rule of thumb above is a good guide for choosing a filter time constant in this filter. If you have a PI-controller, it is advisable to choose the filter time constant proportional to the integral time:

$$T_{\text{filter}} = \frac{T_i}{N}$$

where a rule of thumb is to select N to approximately $N = 10$. This of course assumes that the controller is well tuned with an appropriate integral time T_i.

Integral Windup

Signals in the control loop are always constrained or limited. In particular, measurement sensors have their operating ranges. If the measured variable falls outside the measurement range of the sensor, the process variable is constrained. In the same way, the controller output is constrained. This is illustrated in Figure 3.22.

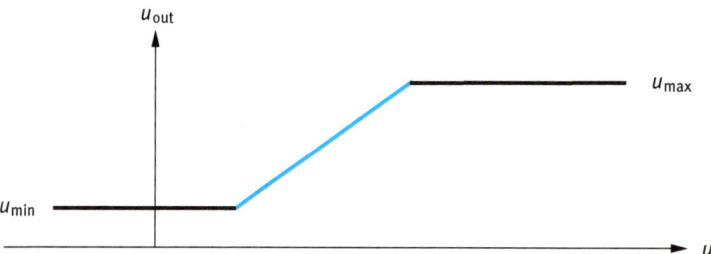

Fig. 3.22: The controller output is in practice always constrained. The figure shows the relationship between the controller output that is set, u, and the controller output that is actually applied, u_{out}.

A valve, for example, has an operating region between fully closed and fully open. The constraints of the controller output can cause problems if the controller does not know when the constraint conditions are met. Figure 3.23 shows the controller output, the process variable and the setpoint in a case where the controller output is constrained. After the first setpoint change, the controller output grows up to its upper limit u_{max}. The controller output is not large enough to eliminate the control error and the process variable stays somewhat below the setpoint.

Earlier in the chapter we showed that the integral part of the PID-controller is proportional to the area formed by the control error. This area is marked in Figure 3.23. As long as there is a control error, the integral part will grow. This in turn means that the desired controller output also grows. We therefore get a difference between the desired controller output u and the controller output as really applied: u_{out}.

Figure 3.23 then shows what happens when the setpoint is decreased to a level where the controller can control the fault.

The change in the setpoint causes the control error to change sign — from positive to negative — which in turn means that the integral part and thus u now begins to decrease. Due to the fact that the integral part grew and became large in the meantime when the controller output had been constrained, the actual controller output u_{out} remains at the limit for a longer period of time. This problem is called integral windup.

The easiest way to fix integral windup is to stop updating the integral part when the controller output is at its constraint or limit. This of course requires that the controller knows what the constraints are. Most process controllers are equipped with methods to avoid integral windup. This is called *anti-windup*.

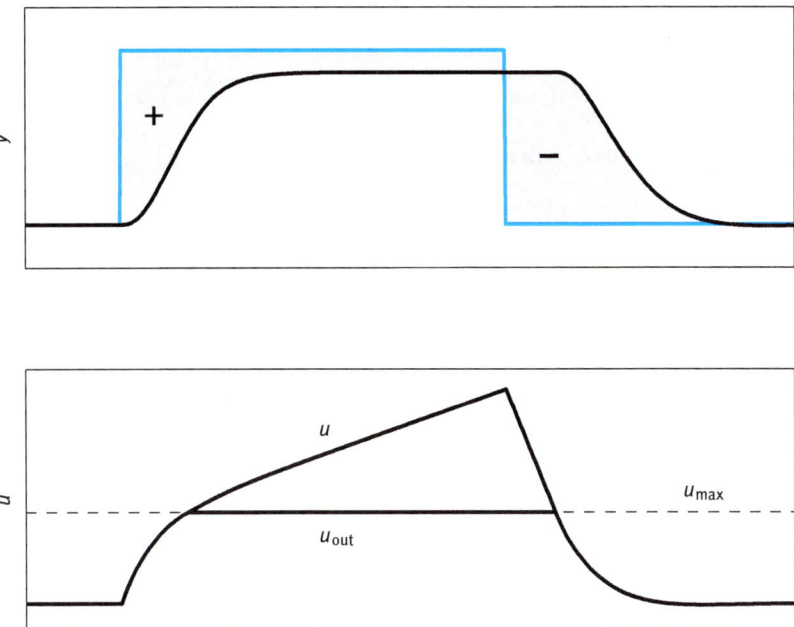

Fig. 3.23: Integral windup with constrained controller output.

Integral windup can also occur in other cases when the controller is unable to eliminate the control error. Examples of such cases are in controllers with selectors and in cascade control. These problems are discussed in more detail in Chapter 6.

Digital Version of the PID-Controller

The PID-controller described so far is analogue, which means that the controller has continuous access to the input signals, that is, setpoint r and process variable y. We also assume that controller output u is continuously updated. This was exactly the case in the old pneumatic and electric implementations but in computer implementations it is no longer true. Instead, in computer implementations the controller input, the process variable, is passed through an analogue-to-digital (AD) converter and the controller output through a digital-to-analogue (DA) converter. This means that the controller output is updated at certain time instances, and kept constant between these instances. The time between the updates is called the sampling period denoted by h.

If you want to implement a PID-controller in a computer, the equations for the analogue PID-controller have to be converted to digital versions. Here, we will show how this conversion can be carried out.

The equation for the analogue PID-controller in parallel form with setpoint weighting is

$$u(t) = K\left(br(t) - y(t) + \frac{1}{T_i} \int e(t)dt - T_d \frac{dy(t)}{dt} \right)$$

The digital control signal can be described as

$$u[k] = P[k] + I[k] + D[k]$$

i.e. a sum of the proportional, integral and derivative parts, where k is the discrete time at which the signal is sampled. The relationship between analogue and discrete time is $t = kh$ with $k = 0, 1, \ldots$. The controller output $u[k]$ is a function of the controller inputs sampled at discrete times: $r[k]$ and $y[k]$.

The proportional part of the analog controller is given by

$$u_P(t) = K(br(t) - y(t))$$

and the digital version simply becomes

$$P[k] = K(br[k] - y[k])$$

The integral part of the analog controller is

$$u_I(t) = \frac{K}{T_i} \int_0^t e(t)dt$$

Here, we have to approximate the integral and do so by replacing it with a sum.

$$I[k] = \frac{K}{T_i} \cdot h \sum_{i=1}^{k} e[i]$$

The approximation is illustrated in Figure 3.24. We can see in the figure that the shorter the sampling period h is, the better the approximation will be. However, the above notation is not a suitable way to implement the integral part, since it would require that all control errors are stored — from the time the controller was switched on. A more useful way to implement the integral part is the recursive form

$$I[k] = I[k-1] + \frac{Kh}{T_i} e[k]$$

This way, only the previous control action $I[k-1]$ needs to be stored.

The derivative part of the analog controller is

$$u_D(t) = -KT_d \frac{dy(t)}{dt}$$

Here we approximate the derivative by replacing it with a difference equation.

$$D[k] = -KT_d \frac{y[k] - y[k-1]}{h}$$

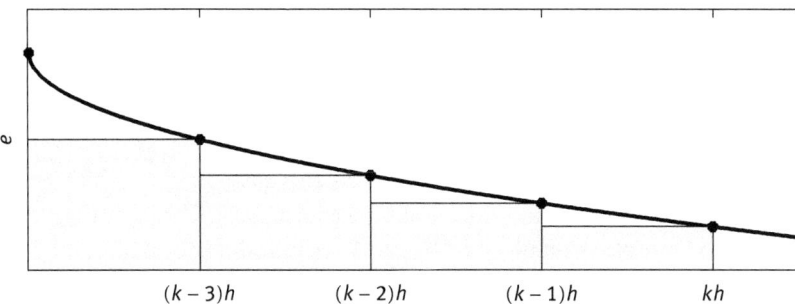

Fig. 3.24: Approximation of the integral of control error e by a sum of rectangles.

The approximation is illustrated in Figure 3.25, and here we can see that the difference approaches the true derivative if the sampling period is small.

One of the reason why the PID-controller is so popular is that it can be implemented with only a few multiplications and difference equations. The digital PID-controller will work almost exactly like its analogue counterpart, assuming that the prerequisite of a small sampling period in comparison to the dominant time constant of the closed-loop system is met.

Most importantly, the equations presented here just show the core of the digital PID-controller. To get a controller that works in practice, aspects such as filtering of the process value, bumpless transfer at mode switches and parameter changes as well as anti-windup, must be added.

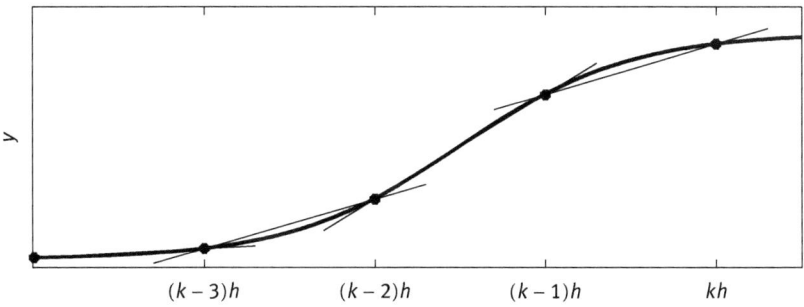

Fig. 3.25: Approximation of the derivative of process output y by straight lines between consequtive values of y.

4 PID-Controller Tuning

4.1 Introduction

The single control loop illustrated in Figure 1.1 consists of two parts: the process to be controlled and the controller that carries out the control task. Chapter 2 discussed the process and the process dynamics. In Chapter 3, we introduced and discussed the structure and function of the PID-controller. The PID-controller has three basic parameters that determine its function, namely the gain K, the integral time T_i and derivative time T_d. These parameters should be selected so that the controller is tuned for the particular process. In this chapter, we will describe different ways to tune a controller.

The chapter begins with a discussion of the control performance we want to achieve, i.e. which problems to solve and what specifications may look like. Then we discuss the most common method of controller tuning, namely manual tuning. In manual tuning, you simply try out successive adjustments of the controller parameters to achieve good performance.

The main part of this chapter is a description of model-based tuning methods. For these methods, the tuning takes place essentially in two stages. The dynamics of the process are first determined, for example with the methods described in Chapter 2. Based on the process description, the controller parameters are calculated according to a formula or to a lookup table. Today, PID-controllers can be tuned automatically. These methods are described at the end of the chapter.

4.2 The Control Task

Before we look at the tuning methods, we will first discuss what we want to achieve, that is, how we want the controller to control the process.

Disturbances

The control loop in Figure 1.1 has three signals: setpoint r, controller output u and process variable y. Figure 4.1 shows the same control loop, but now includes two more important signals: load disturbance v and measurement noise n.

Load disturbances are disturbances that affect the process and interfere with the controller's intention of keeping the process variable close to its setpoint. Examples of load disturbances are uphills, downhills, headwinds and tailwinds when you want to control the speed of a car, solar radiation and variations in outdoor temperature when controlling indoor temperature in homes, disturbance flows when controlling levels and concentrations in tanks and pressure variations in flow controllers.

https://doi.org/10.1515/9783111104959-004

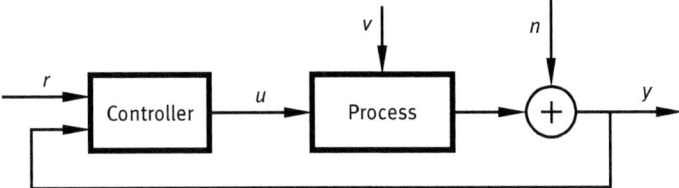

Fig. 4.1: The single control loop indicating load disturbance *v* and measurement noise *n*.

Measurement disturbances are disturbances that do not affect the process itself but the measured process variable. Measurement disturbances are often in the high-frequency range and therefore often called measurement noise. Measurement noise can, for example, be caused by electrical disturbances, vibrations, or quantisation. Measurement noise should be eliminated or filtered out of the process variable, so that the controller tuning and controller performance are not affected by these disturbances. This was described in Section 3.6. Filters will be further detailed in Section 6.2.

Setpoint changes can also be regarded as disturbances in the control loop. However, changes in the setpoint are different from other disturbances because they can be dealt with *before* they are fed to the PID-controller function. Section 3.6 described how the setpoint changes can be handled by the PID-controller using setpoint weighting. To avoid major changes in the controller output due to setpoint changes, you can also filter the setpoint signal or pass it through a ramp function before it is processed by the PID-controller.

In short, we have three types of external signals that affect the control loop: setpoint *r*, load disturbances *v* and measurement noise *n*. Since the measurement noise should be eliminated or filtered out before reaching the PID function and since the setpoint can also be filtered, ramped up and affected by setpoint weighting, load disturbances remain the important signals which should be considered when tuning the controller. Of course, this assumes that there are indeed load disturbances that affect the control loop. In the process industry, load disturbances are very common, while setpoint signals often remain constant for long periods of time.

Speed or Robustness?

The desired control performance of course varies from case to case. Some control loops are important and you would want to spend a considerable amount of time to get them to work well. Other loops may not be that important and you may be content if they work just well enough.

In almost all control loops a compromise needs to be resolved: the requirement for speed versus the requirement for robustness. If you want very fast control, you usually have to accept that the control has overshoots, that means, a concession in robustness. If, on the other hand, very robust control is required, we must accept that the response to setpoint changes and load disturbances will be slow.

Buffer control, which is described in more detail in Section 6.11, is an unusual control problem where we are not primarily interested in the process variable following the setpoint. Instead, you want robust control that leads to a slowly varying controller output. Basically, the only requirement you have for the process variable is that the tank should not be empty or overflowing.

Principle of the Most Common Tuning Methods

Because of the above argument, one can conclude that even if you know the dynamics of the process you cannot have optimal tuning settings of a PID-controller that work in a general setting. Rather, the optimal settings vary from case to case. You can not even say generally when to use a P-, PI-, PD- or PID-controller, although some guidelines are available. This was described in Section 3.3.

The tuning methods described in this chapter are all intended to provide good control in the event of load disturbances.

We still consider single control loops and ignore that other control loops interact with the loop we currently focus on. We will address these problems of interaction in Chapter 6. However, we should point out here that it is important to take into account the surroundings when tuning the controller in a loop. "Interfering" controllers in close proximity should be in manual control during the tuning procedure. Thus, it is possible to distinguish between the effect of the current controller and the effect from other controllers. It is also important to tune controllers in a process section in the correct order. This is also explained in Chapter 6.

4.3 Manual Tuning

Manual tuning usually involves gradually adjusting the controller parameters and studying the effect on the control performance. This is a method that every practically experienced control technician should master. This also applies if you use automatic tuning or any of the model-based tuning methods presented later in this chapter because these methods do not always provide the desired control performance. Therefore, it is often necessary to make manual re-adjustments.

The choice of controller parameters is about finding a compromise between the requirement for fast control and the requirement for robustness. Table 4.1 shows how the speed of response and the robustness changes as you change the parameters.

Note that the table only indicates rules of thumb. There are exceptions. Increasing the controller gain, for example, often results in more robust level control.

Tab. 4.1: Rules of thumb for the control parameter influence on speed and robustness of the closed control loop.

Parameter movement	Speed	Robustness
K increase	increase	decrease
T_i increase	decrease	increase
T_d increase	increase	increase

When tuning the PID-controller manually, you usually tune the parameters in the order P $\rightarrow$ I $\rightarrow$ D, that is, you start with K, then adjust T_i and finally select T_d.

First, T_i is set to a very large value, if it is impossible to switch off the integral part completely, and $T_d = 0$. When K is adjusted so that the control is reasonably good, you decrease T_i gradually until you find a suitable setting. According to Table 4.1, this should result in a deterioration of the robustness, which means that one should reduce gain K slightly.

When the PI-controller provides satisfactory control, one can begin to adjust the derivative part by increasing T_d. According to Table 4.1, this normally means that the robustness is improved, which in turn means that you can increase the gain K and decrease integral time T_i further, thereby increasing the speed of the control response.

According to the table, the derivative part provides for both faster and more robust control when increasing T_d. This is only true up to a certain point. If T_d is selected above a certain threshold, we will have poorer control robustness. As mentioned in Chapter 3, the derivative part estimates the change in the control error at time T_d in the future. This estimate will of course be poor for large choices of T_d.

The trends in Table 4.1 were derived without considering noise and others disturbances. The noise is amplified more as T_d increases. Process noise therefore often is the cause of an upper limit for T_d. In some cases, the noise level can be so high that despite the filtering of the derivative part, you would not want to use the derivative part because it gives a jumpy controller output. A PI-controller is then preferred to achieve a smoother controller output at the cost of poorer control performance.

Tuning Diagrams

The effect of the controller parameters on the control performance described in Table 4.1 is sometimes illustrated in so-called tuning diagrams. Figure 4.2 shows a tuning diagram for a PI-controller. The subplots show how the process variable reacts to a step change in the load disturbance for different selections of the gain K and the integral time T_i. Nine cases are shown, with too large, good and too small values, respectively, for the two parameters.

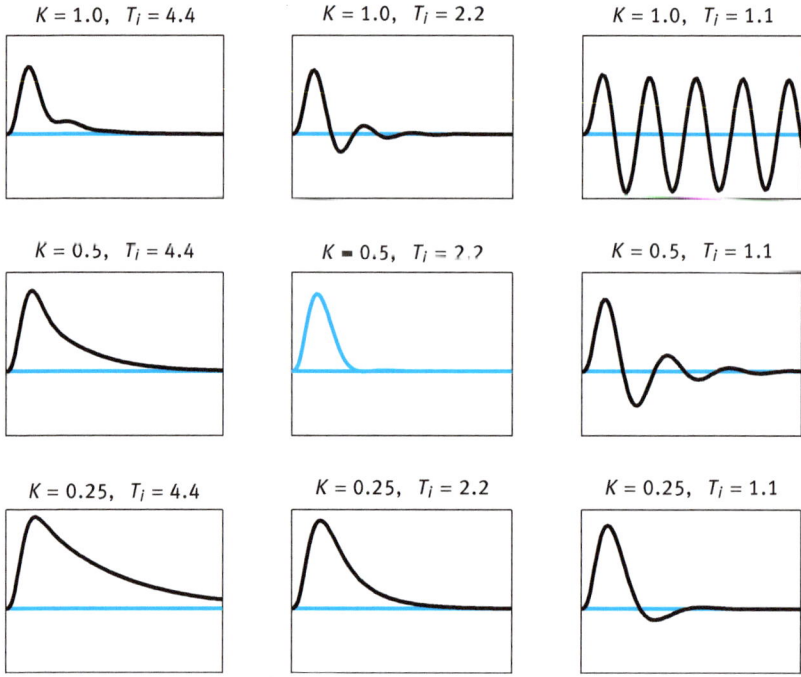

Fig. 4.2: Tuning diagram for the PI-controller. The curves show the response to step changes in the load for different choices of K and T_i. The process is the higher order process of Example 2.2.

As mentioned earlier, you should first achieve a well-tuned PI-controller before tuning a PID-controller. A well-tuned controller corresponds to the middle plot in Figure 4.2 with $K = 0.5$ and $T_i = 2.2$. This controller is tuned so that the area under the control error caused by the load disturbance is minimised, with the additional specification of no overshoot.

When you put a new controller into operation, you usually first get responses that correspond to the example in the lower left plot, that is, the gain is too small and the integral time too long. Unfortunately, experience shows that many, when seeing this, only increase the gain, but do not reduce the integral time. One thus tries to find a good compromise between speed and robustness only by adjusting the gain. As a result, many control loops behave like the upper left plot in the figure where the control is both slow and poorly damped. Normally you want the opposite, that is, fast and well-damped responses.

When a satisfactory control performance has been achieved with the PI-controller, that is, when successfully tuning the control similar to the middle example, you can enable the D-part and to get a faster response while maintaining robustness.

An increase in the derivative time T_d normally results in a more sluggish control, that is, we move down to the lower left plot again. Here is now an opportunity to increase the gain and reduce the integral time so that we come back to the middle plot, but now with a faster response. We can repeat this procedure, that is, increase T_d and then adjust K and T_i, until no further improvement can be achieved.

Summary

As previously mentioned, manual tuning of PID-controllers is an art that every practically experienced control technician should be good at. For a skilled engineer who masters the art, this is most likely the fastest way of getting a control loop to function satisfactorily. An exception are possibly the automatic tuning methods described in Section 4.6.

However, in many cases it may be worthwhile to tackle the problem first by using any of the model-based methods described later in this chapter. Examples of such cases are:

1. Particularly important control loops where we like to spend time to get better control.
2. Particularly difficult loops where we cannot succeed with the normal rules of thumb as our search leads to a dead end. This is because the loop does not respond like most other loops.
3. Particularly slow loops. Because in these cases it takes a long time to test each new set of tuning parameters, it often pays to figure out a good tuning through a model instead of just trying it out.

4.4 Step Response Methods

Model-based tuning methods are based on finding the process dynamics first and creating a model of the process and then tuning the controller based on the characteristics of this model. The most common model-based tuning methods for PID-controllers assume that the process model is obtained through a step response experiment. We will describe several such tuning methods here.

Ziegler-Nichols Method

Ziegler-Nichols tuning rules are among the oldest and most widely used tuning rules. Ziegler and Nichols worked together at Taylor Instruments and published two methods in 1942 for tuning PID-controllers: a heuristic method that is discussed in the next section and a step response method. They developed both methods by conducting a large number of simulations on pneumatic, analogue machines and from these obtained relationships between process models and controller parameters that achieve

Tab. 4.2: PID-parameters of the Ziegler-Nichols step response method.

Controller	K	T_i	T_d
P	$\dfrac{T}{K_p L}$		
PI	$\dfrac{0.9T}{K_p L}$	$3L$	
$\text{PID}_{\text{parallel}}$	$\dfrac{1.2T}{K_p L}$	$2L$	$L/2$
$\text{PID}_{\text{series}}$	$\dfrac{0.6T}{K_p L}$	L	L

well-tuned controllers. To them, a well-tuned controller was a controller that provides a damping of $1/4$, that is, an overshoot that is a quarter of the size of the initial peak. In the process industry, this is often regarded as too aggressive control because you normally want well-damped control loops with zero damping, that is, no overshoot.

The Ziegler-Nichols step response method is based on a step response of the open-loop, uncontrolled process and the model used here was illustrated in Figure 2.5. It provides the controller parameters based on the three values K_p, L and T. This is shown in Table 4.2.

The Ziegler-Nichols table was originally devised for PID-controllers realised in parallel form. We have supplemented the table with the series form using the calculations in Section 3.2. Note that this only applies when we use all three parts of the controller, P, I and D, as this is the only case when there is a difference between the parallel form and the series form.

In the table we see that the controller gain K is inversely proportional to the process gain K_p. This can be easily understood by the following consideration. If the process has a high gain, the controller should compensate for this by having a low gain and vice versa. The integral time T_i and the derivative time T_d are proportional to the dead time L of the process. Note that the controller time unit is of the same order of magnitude as the time unit of the process.

For the PI-controller a lower gain is proposed than for the P-controller. This makes sense because the introduction of the integral part means a deterioration in robustness. Reducing the controller gain compensates for the deterioration of robustness.

For the PID-controller, a gain higher than the gains for both the P-controller and the PI-controller is proposed together with a shorter integral time. This can be achieved because of the damping property of the derivative part.

Unfortunately, the Ziegler-Nichols method gives an unreasonably high gain and an unreasonably short integral time if the dead time is short compared to the time constant. For example, if the process parameters are $K_p = 1$, $T = 600\,\text{s}$ and $L = 1\,\text{s}$, the PI-controller will have a gain that is $K = 540$ and an integral time that is 3 s. Such

Tab. 4.3: PID-parameter of the Ziegler-Nichols step response method for an integrating process.

Controller	K	T_i	T_d
P	$\dfrac{1}{K_v L}$		
PI	$\dfrac{0.9}{K_v L}$	$3L$	
PID$_{parallel}$	$\dfrac{1.2}{K_v L}$	$2L$	$L/2$
PID$_{series}$	$\dfrac{0.6}{K_v L}$	L	L

a high gain is almost always too large and results in a controller that is in practice an on/off controller. This means that the Ziegler-Nichols method cannot be used for this type of process. Such processes are, for example, purely first order or purely integrating processes.

Integrating Processes

The Ziegler-Nichols method can also be used for integrating processes. Section 2.4 described how to derive the speed gain K_v and the dead time L from a step response experiment of an integrating process. Table 4.3 shows how the controller parameters can be determined by using K_v and L. The table shows that the controller gain is inversely proportional to the speed gain K_v, which is reasonable.

It is possible to conduct this experiment even for stable processes. With the definition of time constant T given in Figure 2.5 one can show that the relationship between the speed gain and the process gain is

$$K_v = \frac{K_p}{T}$$

The advantage of this is that you do not have to wait until the process value has stabilised at the new level, but can interrupt the experiment as soon as you have noted the point where the process value grows fastest. This is an advantage especially when the process is very slow. The disadvantage of this latter method is that the determination of the parameters is often not as accurate.

AMIGO Method

The AMIGO method was developed in the early 2000s. AMIGO stands for *Approximate M-constrained Integral Gain Optimisation*. Like the Ziegler-Nichols method, it is based on the study of a large number of simulated processes. One difference is that sixty years after Ziegler-Nichols experiment, computers were used for the simulation instead of pneumatic analogue machines.

The AMIGO method is not only experimental. It also guarantees robustness requirements and it is based on optimisation. The optimisation refers to minimizing the control error when the control loop experiences a step change in the load disturbance.

The AMIGO method is based on first finding the process parameters K_p, L and T from a step response experiment and the 63 percent method as shown in Figure 2.4. The AMIGO rules for a PI-controller are then given by

$$K = \frac{1}{K_p} \left(0.15 + 0.35\frac{T}{L} - \frac{T^2}{(L+T)^2} \right)$$

$$T_i = 0.35L + \frac{13LT^2}{T^2 + 12LT + 7L^2}$$

and the rules for a PID-controller are then given by

$$K = \frac{1}{K_p} \left(0.2 + 0.45\frac{T}{L} \right)$$

$$T_i = \frac{0.4L + 0.8T}{L + 0.1T}L$$

$$T_d = \frac{0.5LT}{0.3L + T}$$

If you want to tune a PID-controller in series form, you can use the conversion formulas described in Section 3.2. Unfortunately, this does not give a result if $0.45 < L/T < 11$ because you can only change from parallel form to serial form if $T_i \geq 4T_d$ and this is not fulfilled in the above interval.

Similar to the Ziegler-Nichols method we see that the controller gain is inversely proportional to the process gain K_p. We also see that integral and derivative times are functions of the process times L and T.

Unfortunately, the AMIGO method has the same shortcoming as the Ziegler-Nichols method when the dead time is significantly shorter than the time constant, namely that it gives an unreasonably high gain and an unreasonably short integral time. For example, if the process parameters are $K_p = 1$, $T = 600\,\text{s}$ and $L = 1\,\text{s}$, the PI-controller will have a gain that is $K = 209$ and an integral time that is $T_i = 13.1\,\text{s}$. This means that the AMIGO method does not work for this type of process, processes that are close to purely first order or purely integrating processes.

The AMIGO rules are a little more complicated than the Ziegler-Nichols rules, but they provide better control in most cases.

Integrating Processes

The AMIGO-method can also be adapted for integrating processes. From the process parameters K_v and L, generated with the methods as described in Section 2.4, the controller parameters for a PI-controller can be computed as

$$K = \frac{0.35}{K_v L}$$

$$T_i = 13.4L$$

and for the PID-controller the AMIGO-method proposes the following parameters:

$$K = \frac{0.45}{K_v L}$$

$$T_i = 8L$$

$$T_d = 0.5L$$

If you want to tune a PID-controller in series form, you can use the conversion formulas described in Section 3.2 and this is always possible since $T_i = 16T_d > 4T_d$.

Lambda Method

The Lambda method was developed in the late sixties and is today widely used in the paper industry. Unlike the methods presented previously, the Lambda method has a parameter that can be adjusted by the user. The parameter corresponds to the dominant control loop time constant. The parameter is called lambda after the Greek letter λ.

The Lambda method is based on first finding the process parameters K_p, L and T from a step response experiment and the 63-percent method as shown in Figure 2.4. The original Lambda method dealt only with PI-controllers and it suggests the following controller parameters

$$K = \frac{1}{K_p} \frac{T}{L + \lambda}$$

$$T_i = T$$

We see here that the integral time is always set to the time constant of the process T. The gain, on the other hand, depends on the choice of λ.

Choosing λ can be difficult. The choice of lambda is a trade-off between speed and robustness, the same balance that one is faced with when tuning the controller manually. A common rule of thumb is to select λ as a factor times the time constant of the process, except in the case where the process has a long dead time. Then you should choose λ as a factor times the dead time instead. The Swedish company SSG Skogsindustriernas Teknik AB gives the following guidelines:

$$\lambda = \begin{cases} T & \text{"Aggressive" control} \\ 2T & \text{"Steady" control} \\ 3T & \text{"Robust" control} \\ 3L & \text{Process with long dead time} \end{cases}$$

Many people think that these guidelines are slightly conservative and prefer therefore the first option, $\lambda = T$, even though this is called 'aggressive'. The resulting rule means that we want the same time constant for the closed loop control as the process has.

An advantage of the Lambda method compared to the Ziegler-Nichols method and the AMIGO method is that it does not provide unreasonably high gain and short integral time for processes with very short dead time L.

As mentioned earlier, the original Lambda method was developed only for the PI-controller, but you can derive equations for the PID-controller similar to the ones for the PI-controller. This can be done in a few different ways depending on the type of process model used. Here is a tuning method that is suitable for the discussed process model:

$$K' = \frac{1}{K_p} \frac{T}{L/2 + \lambda}$$

$$T'_i = T$$

$$T'_d = \frac{L}{2}$$

These rules apply to the series form of the PID-controller. With the conversion formulas presented in Section 3.2 the corresponding rules for the parallel form can be calculated:

$$K = \frac{1}{K_p} \frac{L/2 + T}{L/2 + \lambda}$$

$$T_i = T + L/2$$

$$T_d = \frac{TL}{L + 2T}$$

One-Third Rule

Many controllers in the process industry are poorly tuned or not even tuned at all. It is not uncommon to find loops with factory settings that were entered when the control system was delivered. The reason for this is mostly that the personnel lacks time to tune the controllers. Unfortunately, tuning a controller with any of the model-based methods described above is often considered too time-consuming. The most common method by far for tuning PID-controllers is therefore still the manual tuning of trial-and-error.

The One-third rule is a method for tuning PI-controllers that is simpler than the other tuning methods, but also not as powerful. On the other hand, it is much faster to implement and is therefore useful in cases where there is little time and not too tight requirements for the control performance. The One-third rule can only be used for stable processes and not for integrating processes.

The minimum information needed for tuning a PI-controller for a stable process is a process gain and a time. You need to know the gain of the process to be able to

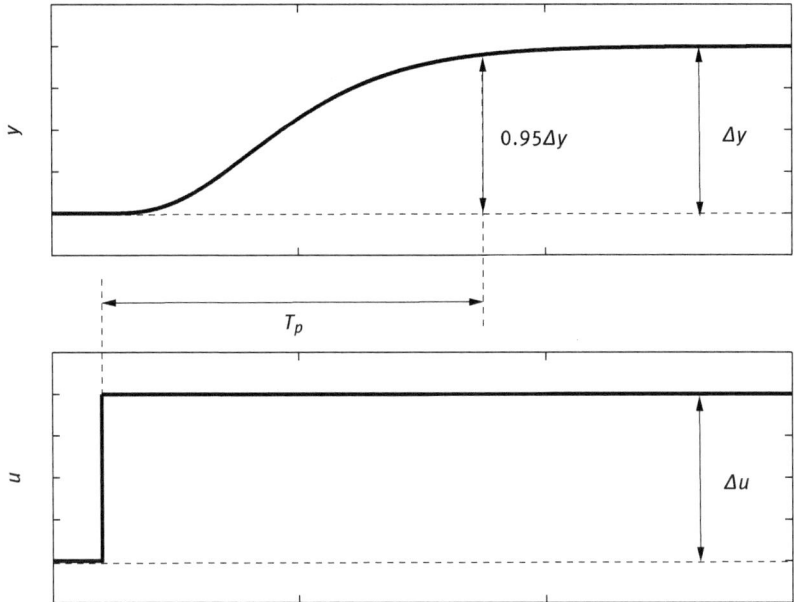

Fig. 4.3: Step response experiment to determine the process parameter using the One-third rule.

choose the controller gain and you need to know how fast the process is to be able to choose the integral time.

The One-third rule is based on performing a step response experiment and determining the process gain K_p and a process time T_p. This is illustrated in Figure 4.3. To use the One-third rule, follow these steps:

1. Write down the current value of the process variable y_1.
2. Do a step change in the controller output of size Δu. Start a timer when you apply the step.
3. When the process variable has almost leveled out, stop the timer and note the process time T_p.
4. Write down the new level of the process variable y_2 and determine $\Delta y = y_2 - y_1$. This will give you the process gain $K_p = \Delta y / \Delta u$.
5. Determine the controller parameters as follows:

$$KK_p = \frac{1}{3}$$
$$T_i / T_p = \frac{1}{3}$$

This procedure can be performed without having to store and plot the process variable and the controller output. It is enough to look at the values displayed on the control system's screen. The expression "almost leveled out" is vague. In Figure 4.3 this time is set to the time it takes to reach 95 % of the final value. However, the method is robust and it is therefore not so important that the time is exactly specified.

The One-third rule can also be used when the process parameters K_p, L and T have been derived from one of the previously described step response methods. In this case, the process time T_p is given by $T_p = L + 3T$ and the tuning rule becomes

$$K = \frac{1}{3K_p}$$

$$T_i = \frac{L}{3} + T$$

There are situations where even the One-third rule is considered too time-consuming. This is often when the measurement range PV_{range} is unknown and difficult to determine. In these cases, it is recommended that the One-third rule be used only to determine the process time T_p and the integral time $T_i = T_p/3$ and that the gain K must be subsequently adjusted manually.

In the same way as the Lambda method, the One-third rule also works for processes with a very short dead time compared to the time constant. In fact, the One-third rule becomes identical with the Lambda method with $\lambda = 3T$ when $L = 0$.

The One-third rule is, as mentioned earlier, designed to be quick and easy to use and remember. It can be briefly summarised as follows:

Perform a step response experiment and set the integral time to one third of the time it takes for the process variable to reach its new level and the gain to one third of the controller output change divided by the process variable change.

Comparisons

To illustrate the properties of the different tuning methods we apply them to the three processes introduced in Examples 2.2 to 2.4.

Example 4.1. Tuning methods for a higher order process

For the higher order process in Example 2.2 the step response analysis gave the following process parameters:

$$K_p = 1 \qquad L = 1.43\,s \qquad T = 2.92\,s$$

The process time T_p, which is required for the One-third rule and is the time it takes for the step response to reach 95 % of the final value, was calculated as $T_p = 7.75\,s$. Table 4.4 shows the controller parameters that result from the methods that were presented earlier.

Figure 4.4 shows the use of a PI-controller when a load disturbance takes place for the following tuning methods: Ziegler-Nichols method, AMIGO-method, Lambda-method with $\lambda = T$ and the One-third rule. The figure indicates that the Ziegler-Nichols

Tab. 4.4: Controller parameters for the higher order processes in Example 4.1. Parameters T_i and T_d are in seconds.

Method	PI		PID		
	K	T_i	K	T_i	T_d
Ziegler-Nichols	1.84	4.29	2.45	2.86	0.715
AMIGO	0.414	2.67	1.12	2.41	0.623
Lambda, $\lambda = T$	0.671	2.92	1.00	3.64	0.574
One-third rule	0.333	2.58			

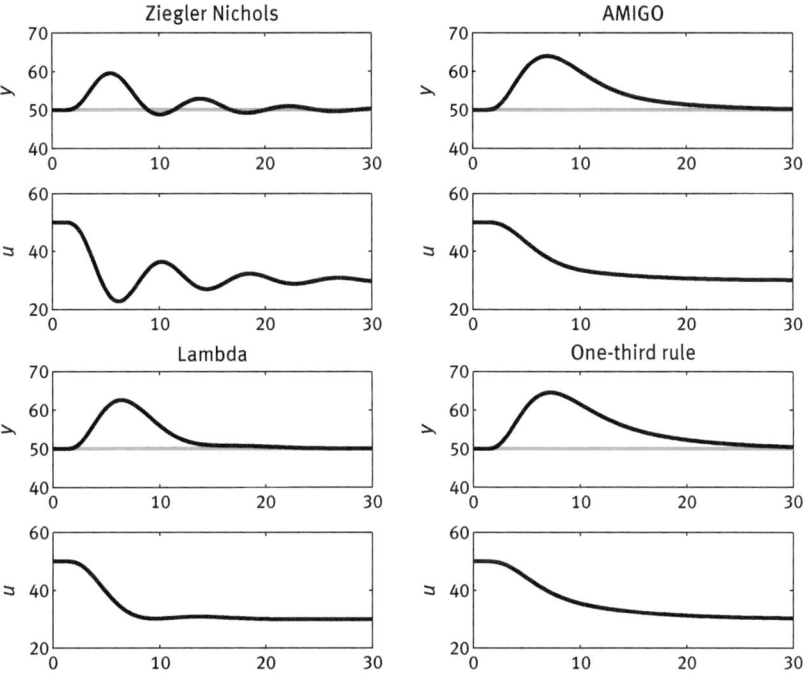

Fig. 4.4: PI-control of a higher order process of Example 4.1. The figure shows the control of a load disturbance at time $t = 1$ s. The controller is tuned with the Ziegler-Nichols method, AMIGO-method, Lambda method with $\lambda = T$ and One-third rule.

provides very aggressive control which results in a small maximum deviation but the trade-off is a resulting oscillation. This type of control is rarely wanted in the process industry. The other methods all provide good control without overshoots. The Lambda method is the method that controls the disturbance the quickest. Figure 4.5 illustrates PID-control in the event of a load disturbance for the Ziegler-Nichols method, the AMIGO method and the Lambda method with $\lambda = T$. The figure highlights that

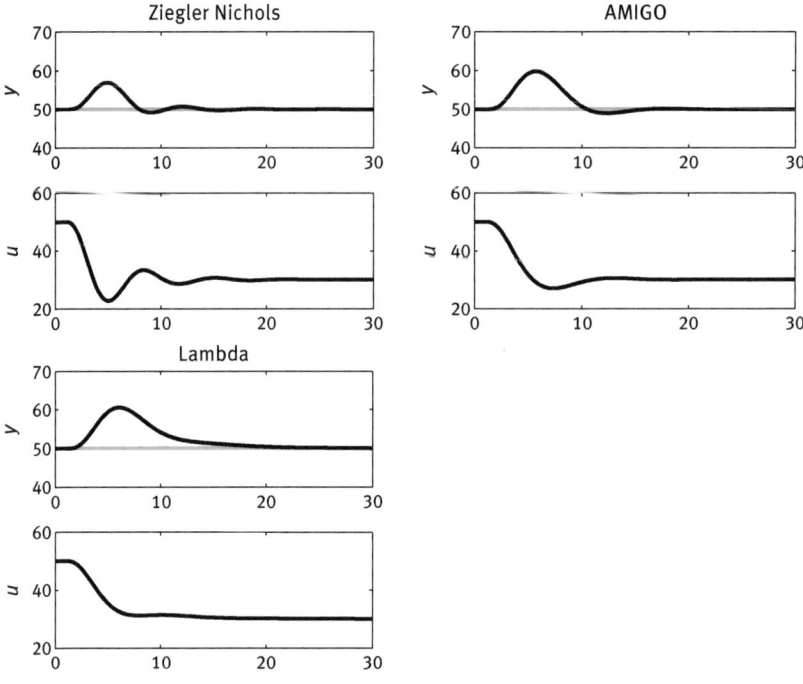

Fig. 4.5: PID-controller of a higher order process as shown in Example 4.1. The figures show control of a load disturbance at time $t = 1$ s. The controllers are tuned with Ziegler-Nichols method, AMIGO-method and Lambda method with $\lambda = T$.

the PID-controller tuned with the Ziegler-Nichols method or the AMIGO method provides significantly more efficient control than the corresponding PI-controllers, while the Lambda method provides approximately the same control for the PI- and PID-controller. Ziegler-Nichols also gives oscillating control, while the AMIGO method and the Lambda method are more robust. □

Example 4.2. Tuning methods for an integrating process
For the integrating process examined in Example 2.3 the step response analysis gave the following process parameters:

$$K_v = 1 \qquad L = 2 \, \text{s}$$

Table 4.5 gives the controller parameters resulting from the Ziegler-Nichols method and the AMIGO-method. The Lambda method and the One-third rule are not listed here because there are no recommendations for integrating processes. Figure 4.6 shows PI- and PID-control with a load disturbance using the two methods. The figure shows that the Ziegler-Nichols method provides a much faster response than the AMIGO method, but at the price of variations in the signals. The figure also illustrates

Tab. 4.5: Controller parameters for an integrating process in Example 4.2. The unit of T_i and T_d is given in seconds.

Method	PI		PID		
	K	T_i	K	T_i	T_d
Ziegler-Nichols	0.45	6.0	0.6	4	1
AMIGO	0.175	26.8	0.225	16	1

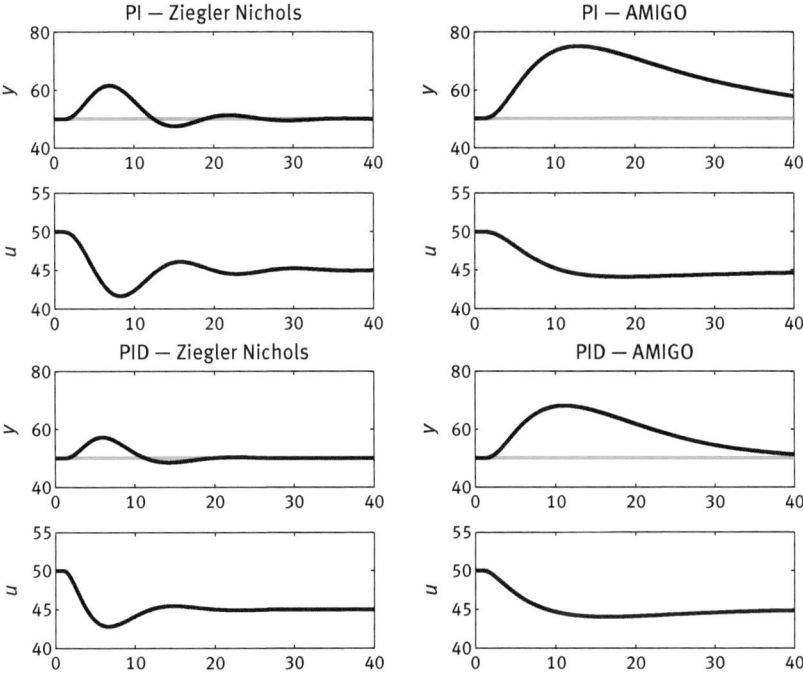

Fig. 4.6: PI- and PID-control of the integrating process in Example 4.2. The figure shows the control of a load disturbance at time $t = 1$ s. The controller is tuned with the Ziegler-Nichols method and the AMIGO-method.

that the PID-controller provides faster control than the PI-controller for both tuning methods. □

Example 4.3. Tuning methods for a dead time process

Processes with long dead times will be dealt with in Chapter 6. Here, we will study how normal tuning methods work for these processes. We described the dead time dominated processes in Example 2.4. The step response analysis of the example gave the following process parameters:

$$K_p = 1 \qquad L = 6.43\,\text{s} \qquad T = 2.92\,\text{s} \tag{4.1}$$

The process time T_p as used in the One-third rule is $T_p = 12.75$ s. Table 4.6 gives the controller parameter derived from the tuning methods that were presented earlier. For the Lambda method we chose the recommendations from SSG for processes with a relatively long dead time, that is, $\lambda = 3L$.

Tab. 4.6: Controller parameter for a dead time dominated process as described in Example 4.3. T_i and T_d are given in seconds.

	PI		PID		
Method	K	T_i	K	T_i	T_d
Ziegler-Nichols	0.409	19.3	0.545	12.9	3.22
AMIGO	0.211	3.61	0.404	4.69	1.94
Lambda $\lambda = 3L$	0.114	2.92	0.273	6.14	1.53
One-third rule	0.333	4.25			

Figure 4.7 shows PI-control with a load disturbance for tuning methods using the Ziegler-Nichols method, the AMIGO method and Lambda method with $\lambda = 3L$ and the One-third rule. The figure highlights that the Ziegler-Nichols method results in a very slow response caused by the large integral time. The other three methods all give well damped and fast responses. The One-third rule is the method that gives the fastest response.

Figure 4.8 illustrates PID-control of a load disturbance of three tuning methods, the Ziegler-Nichols method, AMIGO method and the Lambda method with $\lambda = 3L$. The Ziegler-Nichols method has a relatively high gain and a long integral time which results in a slow and oscillating response. Both the AMIGO method and Lambda method give robust control but the AMIGO method has a faster response.

When comparing the responses of the AMIGO method and the Lambda method with PI-control in Figure 4.7, it appears that the control performance is not much improved. This is an observation which we will return to in Chapter 6 where processes with long dead times are discussed further. □

Conclusions

The step response methods are, as mentioned previously, the most common model-based methods for tuning PID-controllers. Their great advantage is their simplicity. The step change can be easily generated by hand when the controller is in manual mode. The process model can be determined relatively easy if you have the opportunity to record the process variable. If it is not possible to record the process variable, the very simple One-third rule can be used.

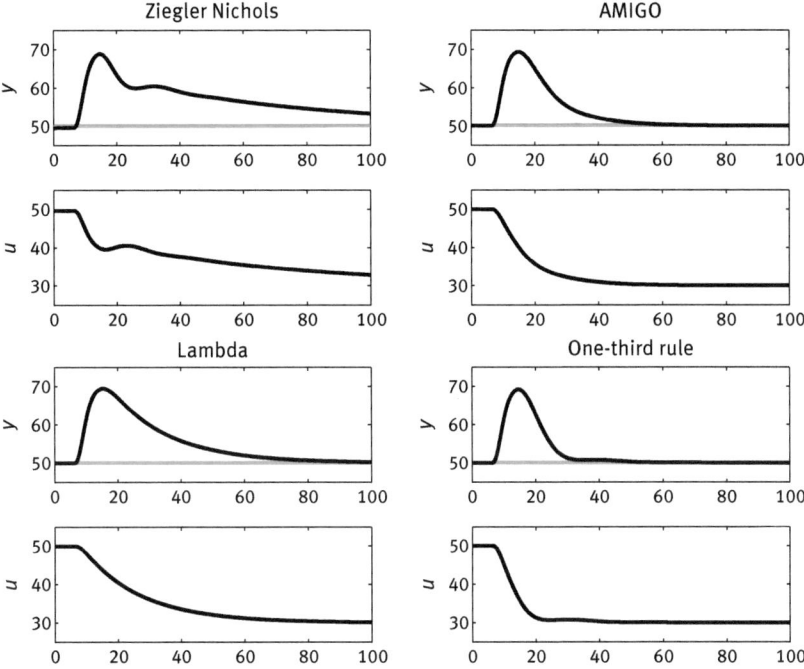

Fig. 4.7: PI-control of the dead time dominated process as described in Example 4.3. The figure shows the control of a load disturbance at time $t = 1$ s. The controller is tuned using the Ziegler-Nichols method, the AMIGO method, Lambda method with $\lambda = 3L$ and the One-third rule.

Unfortunately, the accuracy of these methods is often not very high. This is because you do not want to disrupt the process too much and one is normally forced to do relatively small step changes in the controller output.

As a result, the plots shown in the Figures 2.4 to 2.6 often become small and difficult to read, especially as there also may be measurement noise in the signals. On the other hand, the intention is not to provide perfect control, but rather assist in finding a quick way to get close to reasonably good enough controller tuning. We have previously discussed that the user should always be prepared to make the final adjustments manually, not least because the controller requirements vary from case to case.

There are many methods for tuning PID-controllers based on the step response analysis. We have mentioned four of them in this chapter: the Ziegler-Nichols method, the AMIGO method, the Lambda method and the One-third rule. These are summarised in Table 4.7 for stable processes and Table 4.8 for integrating processes. The Ziegler-Nichols method usually result in too aggressive control, but the others usually give good control for most process types found in the process industry. However, the Ziegler-Nichols method and the AMIGO method do not work when processes have a

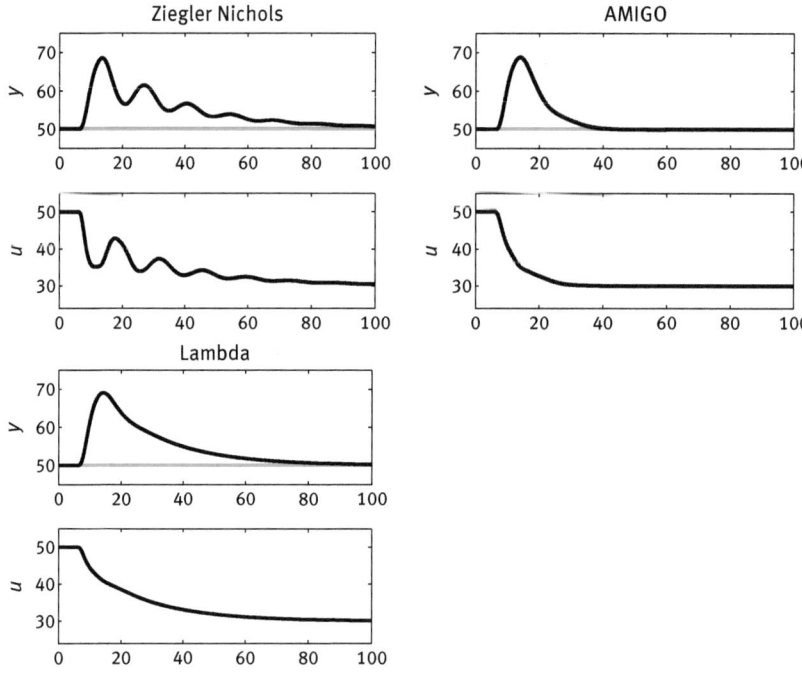

Fig. 4.8: PID-control of a dead time dominated process as given in Example 4.3. The figure shows the control of a load disturbance at time $t = 1$ s. The controllers are tuned with the Ziegler-Nichols metod, AMIGO-method and Lambda method with $\lambda = 3L$.

very short dead time L in relation to the time constant T as they result in a too high gain and too short integral time in these cases.

4.5 Self-Oscillation Methods

Most tuning methods for PID-controllers are based on step response analysis. The step response analysis is simple and convenient, but it has one disadvantage: in certain cases, it can be difficult to determine the process parameters accurately. This may be due to large process disturbances or because the process is so sensitive that the step change in the controller output must be made very small.

There are tuning methods that are based on forcing the control loop to self-oscillate and then determine the controller parameters based on the nature of the self-oscillation. The advantage of these methods is that process dynamics are determined more accurately than with the step response analysis.

Tab. 4.7: Controller parameters based on step response experiments for stable processes: Ziegler-Nichols method (Z-N), AMIGO method, Lambda method and the One-third rule (1/3).

Controller		Z-N	AMIGO	Lambda	1/3
PI	K	$\dfrac{0.9T}{K_p L}$	$\dfrac{1}{K_p}\left(0.15 + 0.35\dfrac{T}{L} - \dfrac{T^2}{(L+T)^2}\right)$	$\dfrac{1}{K_p}\dfrac{T}{L+\lambda}$	$\dfrac{1}{3K_p}$
	T_i	$3L$	$0.35L + \dfrac{13LT^2}{T^2 + 12LT + 7L^2}$	T	$\dfrac{T_p}{3} = \dfrac{L}{3} + T$
PID$_{parallel}$	K	$\dfrac{1.2T}{K_p L}$	$\dfrac{1}{K_p}\left(0.2 + 0.45\dfrac{T}{L}\right)$	$\dfrac{1}{K_p}\dfrac{0.5L + T}{0.5L + \lambda}$	
	T_i	$2L$	$\dfrac{0.4L + 0.8T}{L + 0.1T}L$	$T + L/2$	
	T_d	$0.5L$	$\dfrac{0.5LT}{0.3L + T}$	$\dfrac{TL}{L + 2T}$	
PID$_{series}$	K'	$\dfrac{0.6T}{K_p L}$	See equations in Section 3.2	$\dfrac{1}{K_p}\dfrac{T}{0.5L + \lambda}$	
	T'_i	L	See equations in Section 3.2	T	
	T'_d	L	See equations in Section 3.2	$0.5L$	

Tab. 4.8: Controller parameters based on step response experiments for integrating processes: Ziegler-Nichols method and AMIGO method.

Controller		Ziegler-Nichols	AMIGO
PI	K	$\dfrac{0.9}{K_v L}$	$\dfrac{0.35}{K_v L}$
	T_i	$3L$	$13.4L$
PID$_{parallel}$	K	$\dfrac{1.2}{K_v L}$	$\dfrac{0.45}{K_v L}$
	T_i	$2L$	$8L$
	T_d	$0.5L$	$0.5L$
PID$_{series}$	K'	$\dfrac{0.6}{K_v L}$	See equations in Section 3.2
	T'_i	L	See equations in Section 3.2
	T'_d	L	See equations in Section 3.2

Experiment

Here we describe two experiments that cause the process to self-oscillate. The oscillation can then be used to determine the process dynamics and to derive a process model.

Ziegler-Nichols Method

In 1942, Ziegler and Nichols presented a method for determining the dynamics of the process from a self-oscillation experiment. In the step response methods described in the previous section, the experiments were performed on the process when the controller was in manual control. Ziegler-Nichols self-oscillation method is instead based on the controller being switched on in automatic as a proportional controller. The experiment is performed as follows:

1. Put the controller in automatic mode with integral and derivative parts switched off.
2. Increase the controller gain until the stability limit is reached, that is, until the control loop starts oscillating.
3. Write down the gain the controller has when the loop is self-oscillating. This gain, K_c, is called the critical gain.
4. Measure the time period of the self-oscillation, i.e. the time between two peaks. This time, T_c, is called the critical time period.

Example 4.4. Ziegler-Nichols self-oscillation method

Figure 4.9 shows P-control of the higher order process that was studied in Examples 2.2 and 4.1.

The figure shows that the control loop self-oscillates when the controller gain is $K = K_c = 4$. From the figure the period time of the oscillation can be read out as $T_c = 6.3$ s. The figure also shows control with gains that are larger and smaller than K_c. When the gain is larger than K_c the control loop becomes unstable resulting in growing amplitudes in the signals. When the gain is less than K_c, the control loop becomes stable and the oscillations are damped. □

There are two major problems with the Ziegler-Nichols method. The first problem is that the process lies on the stability limit. This means that you can easily make the control loop unstable if you happen to choose a gain greater than K_c. The second problem is that it is difficult to control the size of the disturbance. The amplitude of the oscillations shown in Figure 4.9 are due to the disturbances in the loop. When carrying out the experiment, one must therefore keep in mind that the process is disturbed just sufficiently so that you get the amplitude oscillating just enough. A major advantage of the Ziegler-Nichols method is that one does not need to know the controller output range OP_range and the process variable range PV_range to determine the process parameters K_c and T_c.

Relay Method

One method that overcomes the disadvantages of the Ziegler-Nichols method is the relay method. The method is also used in methods of automatic controller tuning, described in more detail in the next section. Here we will show how to use the method to manually determine K_c and T_c.

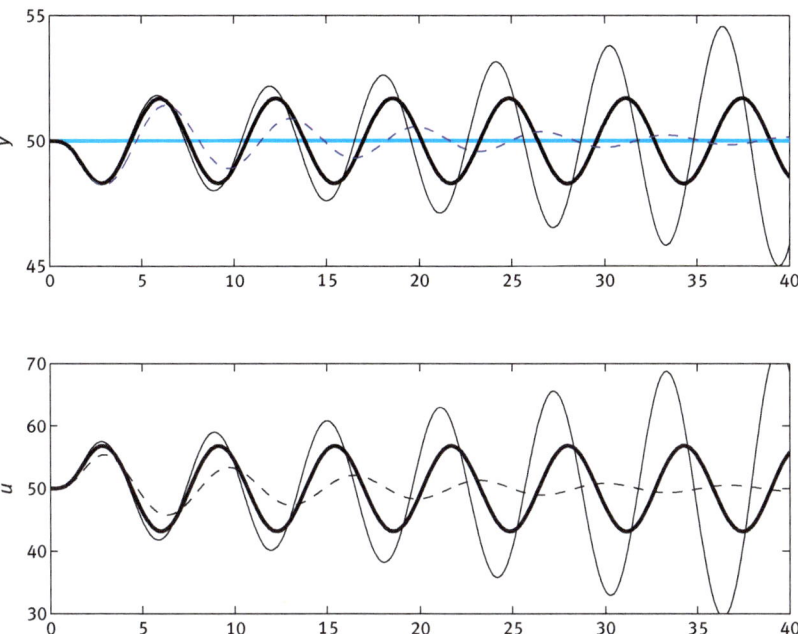

Fig. 4.9: Ziegler-Nichols self-oscillation method for determining K_c and T_c. The thick solid lines show P-control with the gain $K = K_c = 4$. The thin solid lines show control with a higher gain ($K = 4.5$) and the dashed lines control with a lower gain ($K = 3$). In all cases, the loop is disturbed with a short pulse at time $t = 0$ s.

When we introduced the on/off controller in Chapter 3, we found that it had a significant disadvantage, namely that it caused a self-oscillation in the control loop. The interesting aspect is that the oscillation provided by the on/off controller has a time period that is close to the critical time period T_c and that the critical gain K_c can be calculated from the amplitude of the oscillation. The method is called the relay method because the controller output looks as if it had been connected to a relay in the control loop. This is described in more detail in the next section. The relay method works as follows:

1. Leave the controller in manual or automatic control and wait until the process has settled.
2. Adjust the controller output limits so that they are a few percent above and below the current controller output.
3. Put the controller in automatic mode and increase the gain to a very large value, for example $K = 100$.
4. The process variable will now start oscillating around the setpoint. Note the amplitude Δy and time period T_c of the oscillation.

The critical process gain is given by the relationship between the amplitudes of the controller output and the process variable. You have to compensate for the fact that the controller output is a square wave and not a sine wave. One can show that the critical gain is

$$K_c \approx \frac{4\Delta u}{\pi \Delta y}$$

where Δy is the amplitude of the process variable and Δu the amplitude of the controller output. The amplitude Δu is given by the limits we had set for the controller output.

Example 4.5.   Relay method

Figure 4.10 shows the control of the higher order processes studied in Examples 2.2, 4.1 and 4.4 with an on/off controller where the controller output changes between the value $u_{min} = 45\,\%$ and $u_{max} = 55\,\%$.

The process variable oscillates between the value $y_{min} = 48.6\,\%$ and $y_{max} = 51.5\,\%$. This results in the following values for the amplitudes:

$$\Delta u = \frac{u_{max} - u_{min}}{2} = 5\,\% \qquad \Delta y = \frac{y_{max} - y_{min}}{2} = 1.45\,\%$$

This in turn gives us the following estimated critical gain:

$$K_c \approx \frac{4\Delta u}{\pi \Delta y} \approx 4.4$$

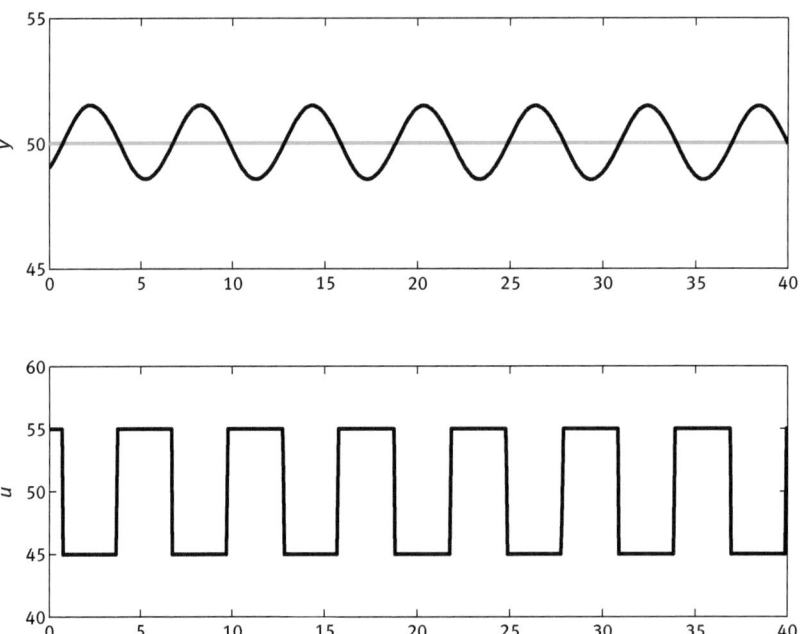

Fig. 4.10: Relay method to determine K_c and T_c.

The time period can be estimated in Figure 4.10 as approximately 6.0 s. In other words, the estimation with the relay method for K_c and T_c lies close to the values identified with the Ziegler-Nichols method in Example 4.4. □

Tuning Rules

There are fewer tuning rules based on the self-oscillation experiments than for step-response experiments. Here we presents two methods: The Ziegler-Nichols method and the AMIGO method.

Ziegler-Nichols Method

Ziegler and Nichols proposed to calculate the controller parameters from the process parameters K_c and T_c as given in Table 4.9. As for the Ziegler-Nichols step response method, we specify the controller parameters for both the parallel realisation and in series realisation of the PID-controller.

Table 4.9 shows that the controller gain is proportional to the critical gain K_c. This is reasonable. For the P-controller, for example, it is recommended to set the gain to half the value determined by the self-oscillation. This corresponds to a gain margin of two. The controller's integral and derivative times are chosen proportionally to the critical time period T_c. This is also reasonable.

Ziegler-Nichols methods are developed with the aim of achieving a damping of 1/4. This is considered usually to be too aggressive in the process industry. Ziegler-Nichols methods do not work very well for processes with long dead times. As mentioned earlier, in these cases it is advisable to use a PI-controller only. However, this should have a lower gain and a shorter integral time than Ziegler-Nichols method provides. A good rule of thumb is

$$K = \frac{K_c}{4}$$
$$T_i = \frac{T_c}{4}$$

This setting provides a lower gain and a shorter integral time than recommended in Table 4.9.

Tab. 4.9: Controller parameters in Ziegler-Nichols self-oscillation method.

Controller	K	T_i	T_d
P	$0.5K_c$		
PI	$0.4K_c$	$0.8T_c$	
PID$_{parallel}$	$0.6K_c$	$0.5T_c$	$0.125T_c$
PID$_{series}$	$0.3K_c$	$0.25T_c$	$0.25T_c$

AMIGO-method

The AMIGO method is also available in a variant based on K_c and T_c. Unlike the Ziegler-Nichols method which is based solely on these two parameters, the AMIGO method also requires the knowledge of the static process gain K_p. The easiest way to determine K_p is to combine the self-oscillation experiment with a step-response experiment. The AMIGO method does not work for integrating processes.

The AMIGO rules for a PI-controller are given as

$$K = 0.16K_c$$

$$T_i = \frac{K_p K_c}{K_p K_c + 4.5} T_c$$

The AMIGO rules for a PID-controller are given as

$$K = \frac{0.3(K_p K_c)^4 - 0.1}{(K_p K_c)^4} K_c$$

$$T_i = \frac{0.6 K_p K_c}{K_p K_c + 2} T_c$$

$$T_d = \frac{0.15(K_p K_c - 1)}{K_p K_c - 0.95} T_c$$

These formulas cannot be translated into the series form because T_i is less than $4T_d$ for almost all processes.

Example 4.6.   Tuning methods for a higher order process

To illustrate the characteristics of the two tuning methods we apply them to the higher order process examined in Examples 2.2 and 4.4. In Example 4.4, the Ziegler-Nichols self-oscillation method gives the following parameters

$$K_c = 4 \qquad T_c = 6.3\,\text{s}$$

In addition, Example 2.2 concluded that the process gain is $K_p = 1$. Table 4.10 gives the controller parameters obtained with Ziegler-Nichols method and with the AMIGO method. Figure 4.11 shows PI- and PID-control with a load disturbance for the two tuning methods. Again, the figure reveals that the Ziegler-Nichols method gives too poorly damped control, but that the AMIGO method works well. The figure also shows that the

Tab. 4.10: Controller parameters of a higher order process in Example 4.6. T_i and T_d are given in seconds.

Method	PI		PID		
	K	T_i	K	T_i	T_d
Ziegler-Nichols	1.60	5.04	2.40	3.15	0.788
AMIGO	0.640	2.96	1.20	2.52	0.930

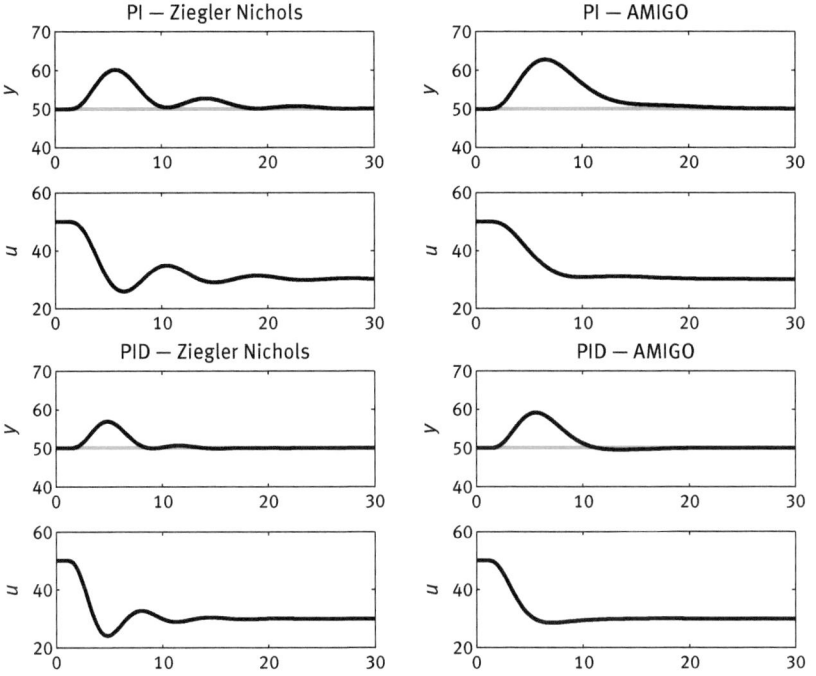

Fig. 4.11: PI- and PID-control of the higher order process in Example 4.6. The figure shows control of a load disturbance at time $t = 1$ s. The controllers are tuned using the Ziegler-Nichols method and the AMIGO method.

PID-controller provides faster control with the same robustness as the PI-controller. A comparison with the results for the step response methods for the same process in Figure 4.4 and Figure 4.5 illustrates that both methods give similarly good control. □

4.6 Automatic Tuning

With the development of computer-based controllers in the 80's there were also opportunities to help the user to tune the controllers. These methods are for the most part fully automated and the technology is therefore called automatic tuning or *auto-tuning*.

In this chapter, we have previously described various systematic methods for tuning PID-controllers. When tuning a controller, you normally go through the following three phases:
1. Introduce a disturbance to determine the process dynamics.
2. Derive a process model by studying how the process reacts to the disturbance.

3. Calculate the controller parameters based on this model.

Automatic tuning of controllers simply means that the above procedure is automated so that the disturbance, the model calculation and the selection of controller parameters take place inside the controller. The work of the engineer is carried out automatically so that instead of determining appropriate controller parameters yourself, you only need to tell the controller that you want it tuned. You possibly need to give the controller some information about the process before the tuning, but in that case this information is significantly easier to specify than the controller parameters themselves.

Auto-tuning of PID-controllers can be done in many different ways. The process can be disturbed in different ways, for example via steps, pulses or by placing the control loop in self-oscillation. The models or process descriptions used may have different forms. Once the model is calculated, there are — as we have seen before — many different ways to choose the controller parameters. As mentioned earlier, this means that you can perform the automatic tuning in many different ways. Here, we explain the principles behind two main types, namely those based on step response analysis and those based on putting the control loop into self-oscillation.

Methods Based on Step Response Analysis

Many methods for automatic controller tuning are based on the process being disturbed by a step change of the controller output. The controller needs to be put into manual mode when giving the instruction to start the controller tuning. Then the controller output is changed in one step to a new value and the controller registers how the process variable reacts to the disturbance. When the process variable has reached its new steady-state position, a model is calculated, usually according to the principles described in Section 2.4. Controller parameters are selected with methods similar to those described in Section 4.4 based on the calculated process model.

The main difficulty when conducting the tuning procedure automatically is to select the size of the step in the controller output. The user of course wants the disturbance to be as small as possible so that the process is not disturbed more than necessary. The auto-tuner, on the other hand, finds it easier to determine the process model if the disturbance is large. The disturbance must be large enough so that the process variable's step response is clearly distinguishable from the noise in the signal.

The noise level can be calculated by observing the variations in the process variable. Even if you know how big the noise level is, you unfortunately do not know how big the step of the controller output should be. For example, suppose you know that the noise level is 2% and therefore you would want the step disturbance to result in a process variable change of 5%. Despite this, one cannot determine how large the step in the controller output should be. To know this, you need to know the process gain K_p.

The result of this dilemma is usually that the user needs to determine how large the step in the controller output should be. Despite this, the auto-tuner is very powerful. The user must only specify the acceptable controller output change instead of finding the controller parameters herself.

Methods Based on Self-Oscillation Analysis

Another possibility is to perform automatic tuning of the PID-controller by determining the process dynamics from self-oscillations as described in the previous section. The biggest problem in step response analysis was to determine the magnitude of the step in the controller output. In self-oscillation analysis one has, of course, corresponding problems, namely to decide the amplitude of the oscillation in the controller output. However, the problem is not as bothersome as for the step response analysis, because you have an opportunity to adjust the amplitude of the oscillation during the experiment. If you note, for example, that the amplitude of the process variable is too large, the amplitude of the controller output can be reduced without interrupting the experiment.

The Ziegler-Nichols method is unfortunately not easy to automate, because it means that you oscillate close to the stability limit. Another disadvantage of the Ziegler-Nichols method is that it is difficult to control the amplitude of the oscillation. The amplitude depends on how much energy is introduced into the control loop. This in turn is due to how much disturbance occurs during the experiment.

The relay method, which was described in the previous section, however, is well suited for automatic tuning. The method is illustrated in Figure 4.12. The controller is in automatic mode when the user gives the instruction to tune the controller. The PID function is temporarily removed and replaced with a non-linear function which can be described as a relay with hysteresis. One then determines the critical gain and the critical time period from the resulting self-oscillation. The controller parameters can then be determined with methods similar to those described in Section 4.5.

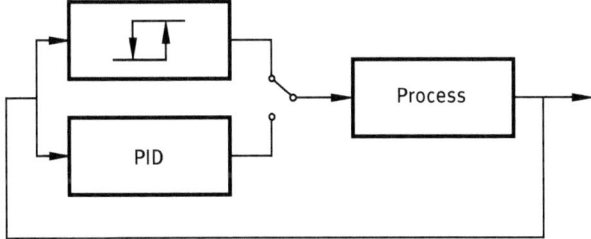

Fig. 4.12: Principle of relay method for auto-tuning.

An automatic tuning of the PID-controller based on the relay method can be performed as follows:

1. First, the noise level in the process is determined by observing the variations in the process variable for a short period.
2. When the noise level is determined, the on/off control is switched on. The hysteresis of the relay is selected based on the noise level. This ensures that the noise does not interfere too much with the experiment.
3. During on/off control, the amplitude of the controller output is adjusted so that the process variable has the desired amplitude. The desired amplitude of the process variable can suitably be selected as a factor times the noise level.
4. After a few periods of oscillation, the time period and amplitude of the process gains can be determined. Suitable controller parameters are calculated based on these parameters.
5. When the parameters have been calculated, the on/off control is switched off and the PID-controller is switched on with the newly calculated parameters.

5 Nonlinear Processes

5.1 Introduction

So far, we have assumed that the process to be controlled is linear. This means that the process response to changes in the controller output is the same for all operating points. The process must also react in a similar way irrespective of whether the controller output increases or decreases. Using the same controller settings for the entire operating range works usually well as long as the process is linear.

In practice, however, there are always nonlinearities in control loops. In many cases you can ignore them, but sometimes the nonlinearities cause problems that need to be addressed. This is particularly true when we have high demands on performance. It is important to remember that by *process* we mean everything that is outside the controller, that is, also sensors, actuators and more. Often, the nonlinearities are found in these peripheral parts of the control loop.

In this chapter, we will describe common types of nonlinearities and give solutions how to deal with them.

5.2 Nonlinear Valves

Nonlinearities in valves are by far the most common cause of unsatisfactorily performing control loops. One reason for this is that the valve is often nonlinear by design, another that additional nonlinearities such as friction and backlash often occur due to wear and lack of maintenance. These problems are discussed in the next sections.

Valve Characteristics

A valve usually has different process gains at different operating points. This can be examined by studying the valve characteristics, that is, the relationship between the signal into the valve and the flow through the valve. Figure 5.1 shows some common types of valve characteristics: Quick-opening, linear and equal percentage. You can modify the valve plug, also referred to as valve cage, so that the valve has an almost arbitrary characteristic, but these three typical cases are the most common ones.

Linear characteristics are the simplest in control terms. If we have a linear characteristic, we can use the same controller parameters for the entire operating range. The quick opening characteristic gives a large flow when you start to open the valve. The equal percentage valve provides a constant relative accuracy and is therefore good if you want to be able to control small flows with great accuracy.

https://doi.org/10.1515/9783111104959-005

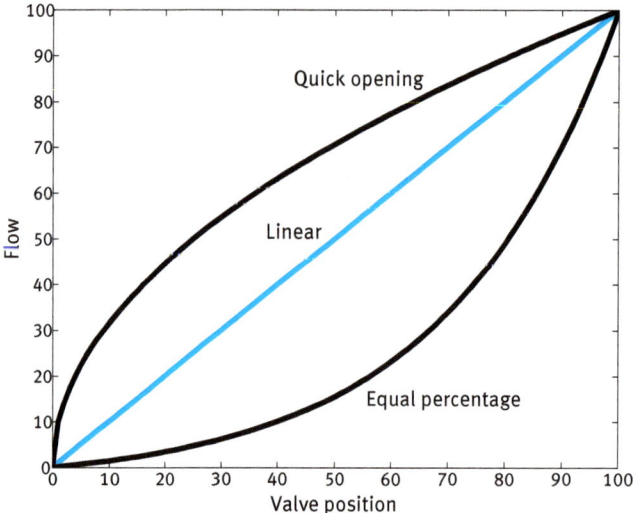

Fig. 5.1: Different types of valve characteristics.

Butterfly valves are by nature quick opening valves, but with the help of a valve positioner you can change the characteristics to linear or equal percentage. In seat valves, the shape of the cone determines the characteristic. By giving the cone a different shape you can get quick-opening, linear or equal-percentage valve characteristics. Here, too, you can change the characteristics via a valve positioner.

Valve specifications and descriptions of the characteristics according to Figure 5.1 are normally made under the assumption that there are no other pressure-reducing elements in the pipeline besides the valve. If you have other throttling devices causing a pressure drop in the pipeline, the valve characteristic will change. To understand this, consider the two cases in Figure 5.2. The flow through the valve is proportional to the square root of the differential pressure, that is, the difference between the pressure before and after the valve. In the upper case, Valve A is the only pressure reducing element, and the valve characteristic will therefore be as expected. In the lower case,

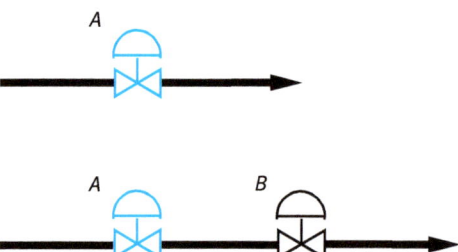

Fig. 5.2: Two cases that give different valve characteristics for valve A.

Valve B reduces the pressure in the line in addition to Valve A. Here, an increase in the valve position of Valve A therefore does not give as large an increase in the flow as in the upper case. The valve is usually not able to give the same flow increase at large valve positions as indicated in Figure 5.1. This means, for example, that a linear valve often has a characteristic similar to that of a quick opening valve when it is installed. In the same way, an equal percentage valve can have a relationship between valve position and flow that looks almost like a linear valve.

From the above reasoning, we can draw the following conclusions about the choice of valve types. A quick-opening valve should only be used for on/off control. A linear valve is preferable if it is the valve that is responsible for the dominant pressure drop in the line. An equal percentage is preferable if there are large pressure losses in the lines outside the valve compared to the pressure reduction across the valve. The equal percentage valve is the most common valve type.

It is difficult to deal with control problems that arise when the valve is nonlinear. For example, suppose we have a quick opening valve. At small valve positions, the flow changes significantly, which means that the valve has a high gain for small valve openings. The quick opening valve, on the other hand, has a low gain for large valve openings. If we tune a controller that will handle the entire valve's operating range, we must give it a low gain so that we can avoid stability problems with small valve openings. However, this leads to slow control at large valve openings, as we then have low gains in both the process and the controller.

If you are not satisfied with the characteristics of the valve, you can solve the problem in different ways. One option is changing the cam disc in the valve positioner, that is, a pure hardware solution. Another method is to compensate for the unwanted valve characteristics in the controller or control system. In many controllers and systems there are linearisation blocks where you can describe the characteristics of the valve by entering a number of points on a curve. With this knowledge, the controller or system can then compensate for the nonlinear characteristic so that the controller perceives it as linear. Another method is to vary the gain of the PID-controller over the operating region in order to compensate for the nonlinear valve. This is called gain scheduling and is described in Section 5.8.

There are many factors that affect the choice of valve, such as the differential pressure, the flow rate, the noise level and the medium. Valves that are oversised are often chosen to be able to safely handle peak loads. Here, the equal percentage profile is a good choice, as it provides a relatively low gain at small valve openings and can also handle large flows at large valve openings.

5.3 Friction

The valve is an exposed component in the control loop. It acts as a brake in the pipeline and is therefore affected by the forces of the transported medium. At the same time, it

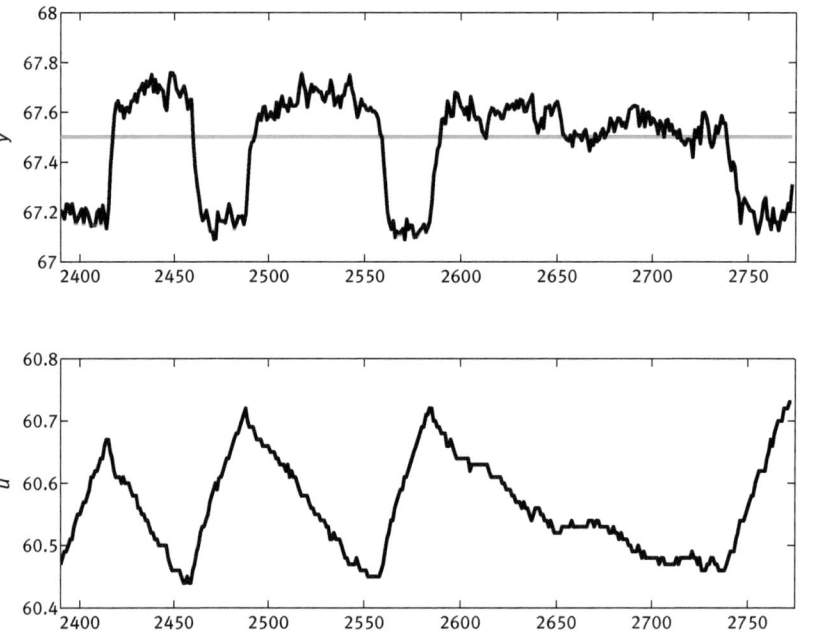

Fig. 5.3: "Stick-slip motion" caused by control with a jamming valve in a flow loop in a distillation column. The time scale is in seconds.

must consist of moving parts in order to vary the flow. These two facts combined mean that the valve is prone to wear. This applies not only to the valve itself, but also to all of its moving parts.

Of course, there is always friction in the movement of a valve. The stuffing box, also referred to as gland packing, allows the valve to move while at the same time preventing the flow from exiting. The problem is that the friction normally increases gradually over the lifetime of the valve. A common reason for this is that the stuffing box of the valve is tightened progressively to avoid leakage. There are several different types of friction. The type that causes the most problems in control valve contexts is static friction, also referred to as "stiction".

A valve with stiction "jams", that is, it does not move smoothly but instead jumps between different positions. This is because a relatively large force is required to overcome the friction in the valve. This force is only reached when the input signal to the valve has changed sufficiently since the previous valve movement. The phenomenon is called *stick-slip motion* or *hunting*.

Figure 5.3 shows the time trend of a control loop with a valve that exhibits a high amount of friction. The figure shows data from a flow controller in a distillation column exhibiting a pronounced oscillatory pattern. While the process variable, the flow, varies much like a square wave, the controller output has a more triangular wave form.

The phenomenon can be explained as follows. Looking at the signals around time $t = 2400\,$s you can notice that the flow is still somewhat below the setpoint, which means that we have a positive control error. The integral part in the controller causes the controller output to grow at a constant rate, which in turn gradually increases the pressure in the actuator. When the pressure has become large enough, the valve plug will jump to a new level, larger than required, where it gets stuck. Here, the flow quickly settles to a new higher level above the setpoint. We now get a negative control error and the controller output starts to decrease and does so until the valve plug jumps back to a new level.

Many people believe that the problem of valve friction is eliminated by using a valve positioner. A valve positioner is a controller that uses a measurement of the position of the valve shaft and controls it to a desired position by applying pressurised air. In fact, the opposite often happens, that is, the valve positioner may cause oscillations in the control loop. The valve positioner can reduce the effect of friction, but does not eliminate it for the following reasons. Firstly, the valve *shaft* may move even though the valve *plug* itself is stuck. Secondly, friction often occurs in the valve controlled by the valve positioner. This friction is often caused by dirt in the instrument air. Many modern valve positioners use PI-controllers, that is, a controller with an integrating part. The integration can cause oscillations in a control loop as explained above, even though the main controller of the control loop is in manual and the valve is expected to be stationary.

When a valve has such high friction that oscillations as shown in Figure 5.3 occur, one should of course carry out maintenance on the valve. Unfortunately, when seeing an oscillating control loop, many people believe that the oscillation is caused by poorly chosen controller parameters and therefore instead lower the gain or increase the integral time in the controller. Fluctuations in control loops can of course be caused by a poorly tuned controller, but this is not very common in the process industry because controllers are usually tuned conservatively. When an oscillatory control loop is detected, one should therefore first determine what the cause of the oscillation is before taking any action.

Figure 5.4 gives guidance for determining the root cause of oscillatory control loops. The first step is to put the controller into manual mode. If the oscillation still persists, we will find the cause of the disturbance outside the control loop. One reason could be that the oscillation is caused by the valve positioner. This can be determined if the valve position is recorded. Because this is often not the case, you simply have to go to the valve and observe if it is oscillating. Another reason for a persistent oscillation may be that an external oscillating disturbance affects the control loop. It is often caused by a nearby control loop that oscillates. In this case, one should try to find the source of the disturbance so that it can be eliminated.

In many cases, the root cause of the disturbance cannot be eliminated. Then all that remains to do is to try to reduce the effect of the disturbance on the control loop. Low-frequency interference is normally not a major problem, because the controller

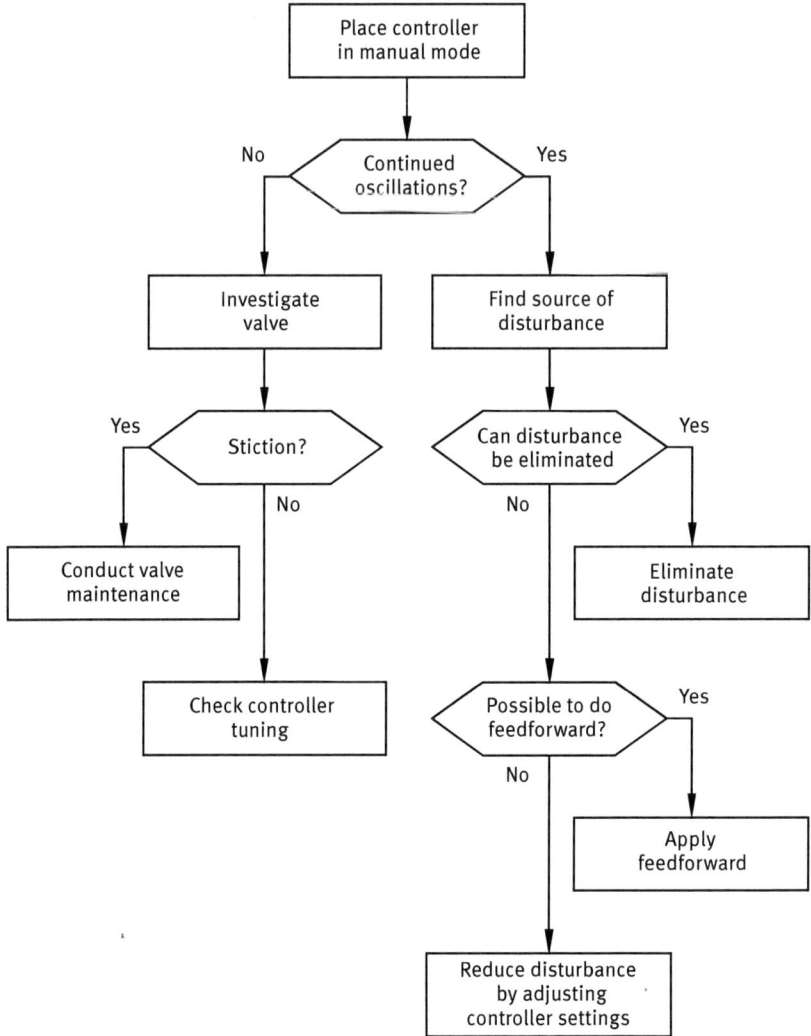

Fig. 5.4: Schema for searching for the cause and correcting oscillations in a control loop.

is able to control these effects. High-frequency disturbances can be dealt with by filtering the process variable so that the disturbances are not fed back to the process. Disturbances that have frequencies close to the critical frequency of the process are problematic. These disturbances are too fast for the controller to eliminate and they are too slow to be filtered out without also filtering out useful process information. If the disturbance or its effect cannot be eliminated you can sometimes use feedforward. This is described in more detail in Section 6.5. If feedforward is not an option, all that

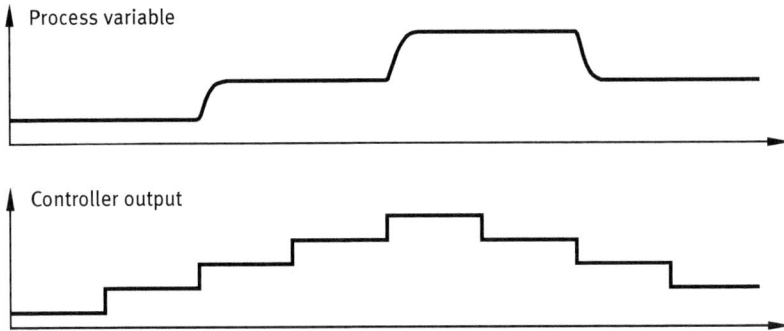

Fig. 5.5: Determination of friction in manual control.

is left to do is to tune the controller parameters so that the effect of the disturbance is as small as possible.

If the oscillations disappeared when we switch to manual control, we can con-clude that the oscillations are generated inside the control loop. Valve friction is the most likely cause. This can be investigated by making small changes to the controller output and checking whether the process variable corresponds to the changes. This is illustrated in Figure 5.5. The process variable only reacts when the controller output has changed sufficiently since the last change in the process variable.

If we find that friction is the cause of the oscillation, the correct action is to main-tain or replace the valve. Often, however, this cannot be done without stopping pro-duction. This normally means that you have to live with the oscillations until the next plant shut-down. There are methods to reduce oscillations caused by stiction. One method is to add a high-frequency signal to the controller output, called a dither sig-nal. Another method is to add small short pulses to the controller output in the change direction of the controller output as you would when you are in the field and stand next to the valve, knocking on the valve so that it gets unstuck. This method is there-fore called "the knocker".

If the valve friction did not cause the oscillation, the controller is the suspected culprit. The correct action is of course to choose better controller parameters when poor tuning is the cause of the oscillation.

Because valve friction is such a common problem in the process industry, there are monitoring methods today that automatically detect oscillating control loops. When an oscillation has been detected, the control loop should be examined in the man-ner described in Figure 5.4. Because friction is everywhere you should always check the valve as shown in Figure 5.5 before re-tuning the controller. If you skip this step you can be easily tricked into tuning incorrect parameters that do not eliminate the disturbance. This also applies when using automatic tuning methods as described in Chapter 4. In addition, you should also check if the valve has backlash, as described in the next section.

5.4 Backlash

A valve that has been subjected to intense wear can cause backlash or hysteresis. Backlash occurs primarily in the actuator when the plug or cage does not move for small changes to the actuator input. The controller output therefore has a "gap" area where it can move freely without affecting the valve. The result is oscillations in the controller output, since the controller output must pass the backlash every time you want to change the direction of the valve movement. The phenomenon is illustrated in Figure 5.6.

The process variable shown in the simulation in Figure 5.6 is relatively unaffected by the backlash. In practice, the process variable also tends to fluctuate, because the gap introduces a dead time to the control loop. In the event of a load disturbance, the backlash must be overcome first before the valve actually moves and starts to compensate for the disturbance.

You can determine the size of the backlash or gap relatively easily by conducting an experiment as illustrated in Figure 5.7. The experiment is performed with the controller in manual mode. First, a step change is made to the controller output, which is so large that the process variable reacts. Since the process variable reacts, we know that we have overcome the gap. Then you make another step change to the controller output in the same direction. Since the step change is made in the same direction, we know that the gap is exceeded during this experiment. Finally, you take an equally big step back, that is, in the opposite direction. We then know that we are doing an experiment in which the controller output must exceed the entire gap first before the valve reacts. The difference between the value of the process variable after the second and the third step response gives us the size of the gap. If we want to calculate the size of the gap in terms of the controller output, we must divide the difference in the process variable by the static gain of the process. In other words, the gap in the controller output, Δu, is given by

$$\Delta u = \frac{\Delta y}{K_p}$$

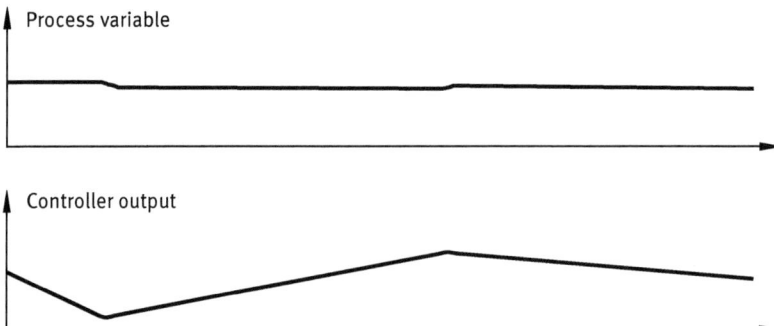

Fig. 5.6: Fluctuations in the controller output caused by backlash in the valve.

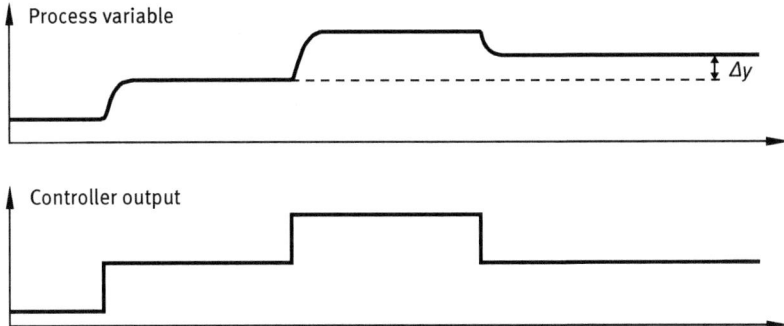

Fig. 5.7: Method for determining the size of the backlash.

where Δy is the difference in the process variable and K_p is the process gain. The process gain can be determined by studying the second step response in the procedure, because here you do a step response experiment in the same direction as the previous one.

When you detect valve backlash, you should of course maintain the valve. Just as with friction, you cannot normally do this without stopping production, which means that in practice you have to live with the gap until the next shutdown. There are some approaches to compensate for valve backlash through the controller. If you know the size of the gap, you can let the controller output "jump" over the gap when you want to change the direction of the valve movement. However, these methods are still rare in industrial controllers.

We have previously said that it is important to check whether the valve has excessive friction before trying to tune the controller. This also applies to backlash. Most methods for automatic controller tuning underestimate the process gain if the gap is large. In these cases the controller gain is then often tuned too high. Finally, it should be noted that valves often have both too high friction and backlash.

5.5 Nonlinear Sensors

Sensors are often nonlinear. The reason for this is that the transmitter often does not measure the physical quantity we are interested in directly, but another quantity that can be related to it. We will give examples of this here.

Temperature Measurement

There are many different ways to measure temperatures. For example, resistance thermometers, such as Pt100 sensors, make use of the fact that the resistance of certain materials varies with temperature. By sending a current through such a material, usually a platinum wire, and measuring the ratio between voltage and current, you can determine the resistance and thereby indirectly the temperature. However, the relationship between resistance and temperature is nonlinear. Therefore, the signal must be compensated for this nonlinearity. This is done either in the sensor or in the controller.

Thermocouples make use of the following phenomena. Two metals of different composition connected at one point generate an electrical voltage that varies with the temperature difference between the metals. Again, the relationship between temperature and voltage is nonlinear. However, many sensors and controllers are equipped with calibration tables so that these signals can be linearised. In this way, the controller can use a process variable proportional to the temperature.

Flow Measurement

Flows are often measured using differential pressure sensors. The differential pressure is a nonlinear function of the flow because the pressure is proportional to the flow squared. To get a linear process, you should therefore take the square root of the process variable before the measurement is used by the PID function. This is illustrated in Figure 5.8. The sensors are often equipped with a square root calculation, which means that the signal from the sensor is already linear. In order to deal with cases where the sensors do not linearise the signal themselves, most process controllers allow you to take the square root of a process variable.

Level and Volume Measurement

Level measurements in a tank are also often done using pressure sensors. This principles utilises the fact that the pressure at the bottom of the tank is proportional to the liquid level multiplied by the density of the liquid. Here we immediately see a problem, namely that the pressure not only varies with the level but also with the density of the liquid. If the density varies, this will in turn affect the accuracy of the level measurement.

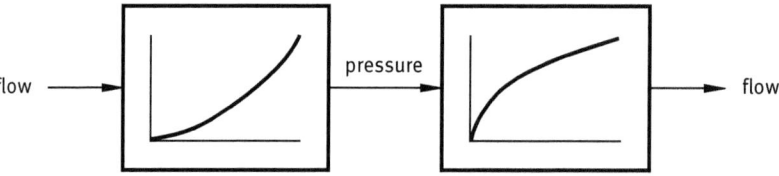

Fig. 5.8: Compensating the signal from a differential pressure sensor to provide a flow signal.

When measuring the level in closed tanks, it is not enough to measure the pressure at the bottom of the tank. You must measure the pressure at two levels to determine the level in the tank through the pressure difference.

Often you do not want to measure the level in the tank but the volume. This is easy when the tank consists of vertical walls, but if the tank for example is spherical, conical or has the shape of a horizontal cylinder, the volume varies with the level. In these cases, one should compensate for the nonlinearity of the volume by using equations that relate the level to the volume.

5.6 pH-Control

pH control is common in neutralisation processes. For example, you often want to neutralise an acidic sewage flow before discharging it. This is done by adding an alkaline substance. Of course, the reverse is also true: you can neutralise an alkaline flow by adding acid. Sometimes the flow can be alternately acidic and alkaline. In these cases, it must be possible to add both an acidic and an alkaline agent. A control strategy to solve this problem is split-range control, described in Section 6.7.

Neutralisation can take place in two ways. Either the neutralising medium is added directly via a mixing valve to the pipeline that contains the flow to be neutralised. However, the control task is often simplified by letting both flows mix in a tank so that fast variations in the input pH value are less pronounced.

The relationship between the concentration of the neutralising medium and the resulting pH-value is often highly nonlinear. The relationship can, for example, look as shown in Figure 5.9. Here, we clearly see how the process gain varies between different operating points: a large gain for very small and very large concentrations is interspersed with a very low gain.

If you want to control the nonlinear process with a controller with constant parameters, these parameters must be tuned for the worst case. This means that we must give the controller such a low gain that we do not get stability problems when the process

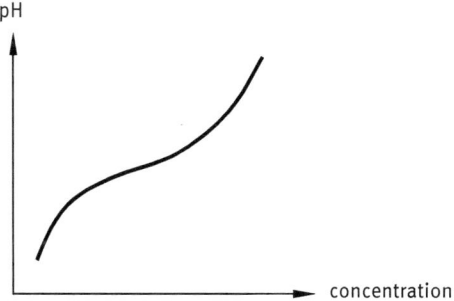

Fig. 5.9: Example of relationship between concentration of neutralizing medium and pH value.

is at its highest gain. The result is that we get slow control in the situations when the process gain is low.

One solution to the problem is to tell the controller what the nonlinearity looks like. Then the controller can adjust its parameters according to the gain of the process for the current operating point. One way to do this is to use gain scheduling as described in Section 5.8.

pH control is one of the most difficult control problems in the process industry. The reason are nonlinearities as the one depicted in Figure 5.9. The problem is often further complicated by the fact that the nonlinearity varies. To successfully control the process, one is often forced to measure the feed from incoming flows and concentrations. This is described in more detail in Section 6.5. pH-control is also an example where adaptive control, described in Section 5.9, has shown to improve control performance.

5.7 Asymmetrical Processes

Most heating processes, for example furnaces, are nonlinear processes because the dynamics vary depending on whether the controller output increases or decreases. In these cases, we can raise the temperature relatively quickly by adding more energy. However, it takes much longer to lower the temperature in the oven because the oven is insulated and we usually do not have any active cooling. In technical control terms, this means that the process has a short time constant when the controller output increases and a long time constant when the controller output decreases. This is illustrated in Figure 5.10.

The best way to compensate for this nonlinearity is to give the controller different parameters depending on the change of the controller output; low gain and short integral time when increasing the controller output and high gain and long integral time when decreasing the controller output. Otherwise, we have to accommodate for the worst case, that is, let the controller operate with low gain and long integral time in both cases.

Most industrial PID-controllers do not allow to change the controller parameters according to the change in direction of the controller output. However, gain scheduling, described in the next section, allows changing the control parameters depending on the controller output.

5.8 Gain Scheduling

So far we have described different types of nonlinearities. In cases of friction and backlash, the best course of action is to maintain or replace the valve. In other cases, you can compensate for the process nonlinearity by letting the controller counteract it.

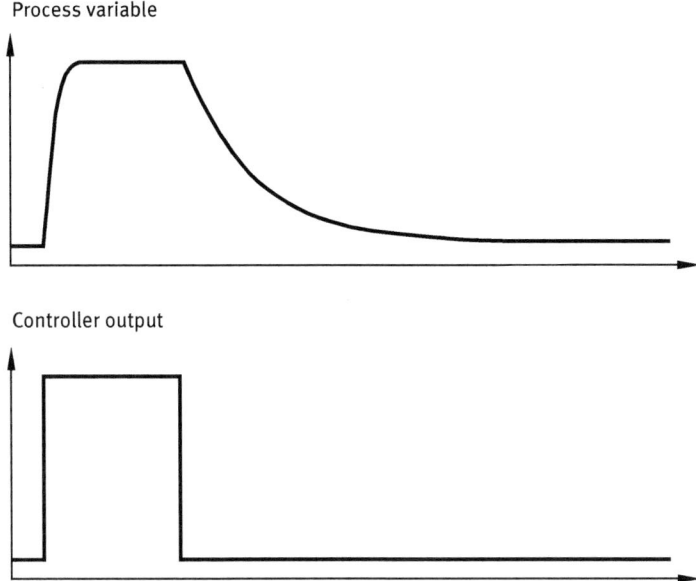

Fig. 5.10: Step response of an asymmetrical process.

Sometimes, when you know exactly what the nonlinearity looks like, you compensate very accurately. Examples of this are linearisation of thermocouples and taking the square root of the differential pressure measurement to deduce the flow rate. Unfortunately, in most cases we cannot compensate for the nonlinearities as easily. Neither is it necessary that the compensation is very accurate. In this section we will show how you can compensate for nonlinearities with the help of gain scheduling.

Assume that we have a valve with an equal percentage characteristic as shown in Figure 5.1. If we control the process with only one set of controller parameters, we must choose the parameters so that we can handle the worst case, that is, the high process gain at large valve positions. This works, but as explained previously, gives a slow control at small valve positions.

The principle for gain scheduling is to divide the controller's operating range into several smaller regions and use different controller parameters in the different regions. This means that you construct a table with different sets of controller parameters as shown in Table 5.1.

In Table 5.1 there are three sets of controller parameters. You can of course have more sets if you want more accurate control. Gain scheduling requires access to a signal that tells you when to switch between the parameter sets. This signal is called the gain scheduling reference or GS_{ref}, in Table 5.1. In the case presented here, the reference signal is the controller output. The figure shows that the controller gain is 0.43 for controller outputs in the range 0–33 %, 0.28 for controller outputs in the range 33–67 % and 0.20 for controller outputs above 67 %. This gradual decrease of the con-

Tab. 5.1: Table with controller parameters for gain scheduling for a valve with equal percentage characteristic.

GS_{ref} [%]	K	T_i	T_d
0 – 33	0.43	1.81	0.45
33 – 67	0.28	1.89	0.47
67 – 100	0.20	1.71	0.43

troller gain compensates for the valve's gradual increase of the process gain for increasing valve positions.

Gain scheduling is a very simple and useful method. Two things are required for it to work. First, you need to find a suitable reference signal, GS_{ref}, which tells you when to change controller parameters. This choice is described for some examples below. Secondly, it is necessary to divide the operating range in an appropriate way. It is usually not a good choice to divide the operating range into equal parts, as is done in Table 5.1. It is common to want a finer division in some areas and to accept a coarser division in others. The division depends on the shape of the nonlinearity.

Example 5.1. Gain scheduling for nonlinear valves

Figure 5.11 shows the principle of gain scheduling applied to a flow controller with a nonlinear valve.

In this case, the valve position determines where we are on the valve characteristics, or — in other words — the valve position determines the process gain. The signal that tells best which valve position we are currently at is the controller output. The controller output should therefore be used as a reference signal for gain scheduling. □

Example 5.2. Gain scheduling for level control of nonlinear tank

Figure 5.12 shows the principle of gain scheduling when controlling the level in a tank with a varying cross-sectional area. The process gain varies depending on the level because of the shape of the tank. At low levels we have a small cross-sectional area in the tank. This means that the level changes quickly if we vary the flow into the tank.

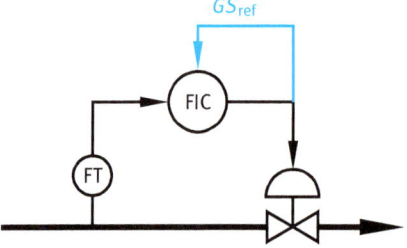

Fig. 5.11: Gain scheduling to compensate for a nonlinear valve. FIC, *Flow Indicating Controller*, is a flow controller and FT, short for *Flow Transmitter*, is a flow sensor.

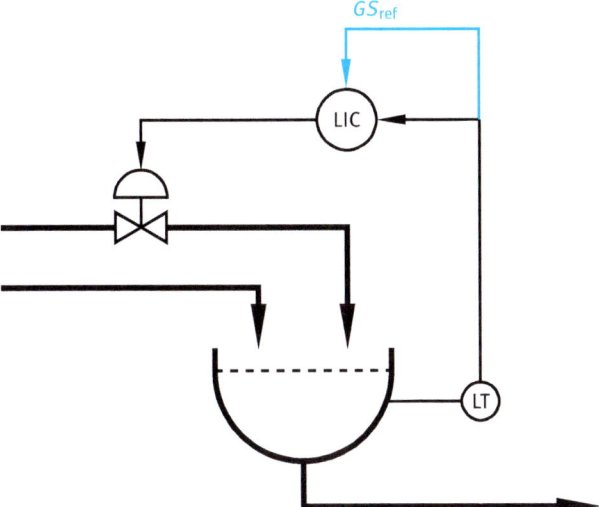

Fig. 5.12: Gain scheduling for level control of a nonlinear tank. LIC, short for *Level Indicating Controller*, is a level controller and LT, *Level Transmitter*, is a level sensor.

In other words, the process has a *high* gain when the level in the tank is *low*. At high tank levels, we have a relatively large cross-sectional area in the tank. The process is therefore slower and in this case has a lower gain.

In this example, it is beneficial to have different controller parameters at different tank levels. Gain scheduling therefore depends on the level, that is, the process variable. The controller output, which determines the controlled flow into the tank cannot be used as a reference signal in this case, since it is not related to the tank level at all and is thus unrelated to the nonlinearity. This is especially easy to see if you have an additional disturbance flow into the tank, as indicated in the figure. □

Example 5.3. Gain scheduling when controlling temperature for heat exchangers
Figure 5.13 shows the principle of gain scheduling for temperature control for a heat exchanger. The temperature controller controls the valve on the primary side based on the temperature on the secondary side. The process dynamics change when the flow on the secondary side varies. For example, if the flow on the secondary side decreases, the liquid on the secondary side will be in the heat exchanger for a longer period of time and thus be heated for a longer period of time. This means that the process gain increases.

In this case, neither the controller output (valve position) nor the process variable (temperature) gives information on the changing process dynamics. Instead, the flow on the secondary side does. By constructing a gain scheduling table with the flow on the secondary side as a reference signal, we can compensate for the variations in the process dynamics. □

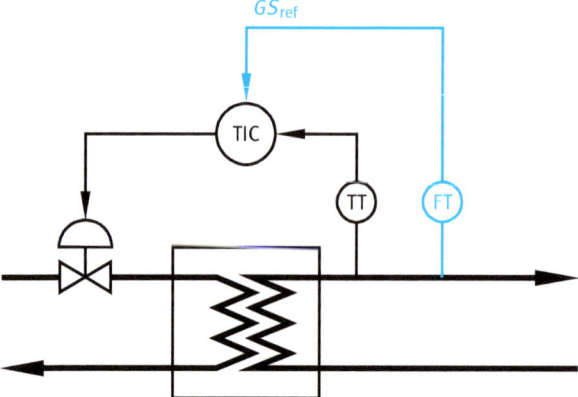

Fig. 5.13: Gain scheduling for temperature control for heat exchangers. TIC, *Temperature Indicating Controller*, is a temperature controller, TT, *Temperature Transmitter*, is a temperature sensor and FT, *Flow Transmitter*, is a flow sensor.

The three examples show that it is crucial to choose the right reference signal for gain scheduling. The reference signal should be the signal that best describes where we are on the nonlinear operating range. We have the following three categories to choose from:

1. Nonlinear actuators such as valves and pumps: The controller output should be used as the reference signal for gain scheduling. This also applies to split-range control described in Section 6.7.
2. Nonlinear sensors: The process variable may be used as the reference signal for gain scheduling.
3. Production-dependent nonlinearities: Any signal that is related to the production may be used as the reference signal for gain scheduling. This can be the process variable, but also some external signal that may be passed to the controller. For the asymmetric heating process described in Section 5.7, for example, the change in the controller output determines which parameters we should use. We can generate this signal by sending the controller output through a function block that calculates the signal's change of direction, that is, whether the controller output is increasing or decreasing.

To use gain scheduling, it should be emphasised that you do not need to know exactly what causes the nonlinearity, even if this knowledge is an advantage. If we know, for example, that a control loop works differently at high production rates compared to low production rates, we can use a production signal as a reference signal for the gain scheduling and switch between two sets of controller parameters.

There are cases for which you may need more than one gain scheduling reference signal. We can, for example, have both a nonlinear valve and production-dependent variations. Some control systems allow to implement this.

Gain scheduling is an efficient, robust and simple control strategy. Aircrafts have long been controlled with the help of gain scheduling. An aircraft is a nonlinear process that has completely different dynamic properties at low altitude and speed, for example during take-off and landing, compared to high altitude and high speeds. In an aircraft, the controller parameters are varied both in terms of altitude and speed.

Gain scheduling can also be used to vary the control performance for different operating ranges. Section 6.11 describes buffer control. There you want a robust control for most of the operating range, while you want an aggressive control when the level in the buffer tank approaches the high or low limit. In some cases, you can also improve the control performance by using aggressive control in case of large control errors, while allowing the control to be more robust in case of small control errors.

Gain scheduling has not been used much in the process industry until recently. There are several reasons for this. Firstly, it only became practically feasible to implement gain scheduling with the introduction of computer-based controllers. Another aspect that has contributed to gain scheduling being quite common today is automatic tuning of controllers. Gain scheduling means that you have to construct a table with several sets of controller parameters. This is a time-consuming job if you do not have access to automatic tuning.

5.9 Adaptive Control

Another method of compensating for nonlinearities is to use adaptive control. The academic world paid a lot of attention to adaptive control in the nineteen sixties and seventies. To adapt means to readjust. Adaptive control means control in which the controller has the ability to adapt to changing process dynamics. The principle of adaptive control is shown in the Figure 5.14. In addition to the usual controller parts, adaptive control has an "adapting" control loop. The adaptive controller continuously monitors the control performance by observing the controller output u and the resulting process variable y. By studying these signals, the controller can calculate a model of the process dynamics. Based on this model and based on a suitable tuning method, the adaptive controller can then decide which controller parameters to use.

The advantage of adaptive control is the ability to detect and adapt to variations in the process dynamics. Assume, for example, that the process gain for some reason suddenly increases. The adaptive controller will see that the controller output results in a different process behaviour compared to the previous behaviour. In other words, the process model will — after a while — describe the process with its new, larger gain. Thus, the controller gain is decreased. In this way, the adaptive controller automatically compensates for the variations in the process dynamics.

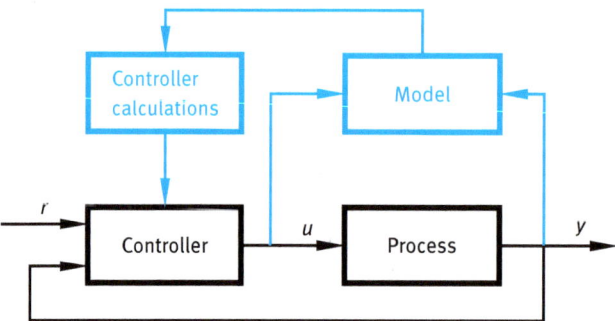

Fig. 5.14: Principle of adaptive control.

So why do we not replace all controllers with adaptive controllers, if they have the benefit of following dynamic variations in the process, if such should occur? There are several reasons for this.

The main reason is that an adaptive controller is more sensitive to disturbances than a controller with constant parameters. The adaptive controller will interpret a disturbance with high energy in the process's interesting frequency as a change in the process dynamics.

Assume, for example, that the control loop oscillates due to friction in the valve. This was discussed in Section 5.3. The adaptive controller has difficulty determining whether the oscillation is due to friction in the valve, an external disturbance, or whether it is caused by the controller being poorly tuned. The result is usually that the adaptive controller reduces the controller gain to try to get rid of the oscillation. In other words, a controller with constant parameters is more robust against disturbances than an adaptive controller.

Areas of Application for Adaptive Technology

We have now gone through two methods to compensate for nonlinearities: gain scheduling and adaptive control. We will end this chapter by discussing when it is appropriate to use these two methods.

First and foremost, it has to be said that almost all control problems can be solved with PI-controllers with constant parameters, provided that there are relaxed requirements for the control performance. However, if you have greater demands on the control performance you sometimes have to choose other structures for the controller.

If the process is nonlinear, you can manage with a PI-controller by setting the controller to accommodate the worst case. If you want better control, however, you must tell the controller that the process is nonlinear, for example by using gain scheduling or adaptive control.

The selection of the controller structures is described in the Figure 5.15. The figure shows how the dynamics of the process may determine which controller structure should be used when there are high demands on the control performance.

If the process dynamics do not vary, there is no reason to have a controller with varying dynamics. If the process has the same dynamics in all operating areas and cases, you can in other words manage with a controller that is set once and for all with a suitable set of controller parameters.

If the process dynamics vary, the controller should compensate for this by letting its parameters vary. One can distinguish between two types of dynamic variations, namely those that are predictable and those that are unpredictable. By predictable dynamic variations, we mean dynamic variations that we can relate to a signal in our control system.

The best way to handle processes with predictable dynamic variations is to use gain scheduling. One could of course imagine solving the problems with the predictable dynamic variations by using an adaptive controller. The advantage of this is that you then do not have to fill in the gain scheduling table and to specify a reference signal.

The advantage of gain scheduling compared to adaptive control is first and foremost that it is more robust. Another advantage is that you tell the controller *how* and *when* the dynamics vary. In adaptive control, we only know *that* the dynamics of the

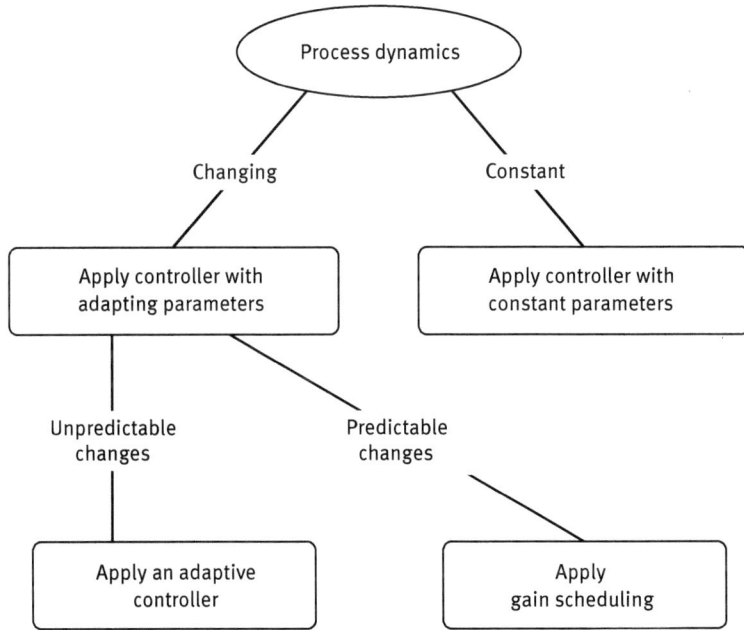

Fig. 5.15: Selection of controller structure.

process can vary, but not *how* and *when*. The difference means that an adaptive controller always needs time before it discovers what dynamics the process has at the moment. It will thus take time to find good controller parameters.

Assume, for example, that there is a nonlinear valve in the control loop. In gain scheduling, the controller is told what the valve characteristics look like and can thus always work with a suitable controller gain. If, on the other hand, you use adaptive control that has found the right gain at a certain operating point and then changes the operating point, it will need a certain amount of time before it has adapted to the new gain.

If the dynamic variations of the process are not predictable, gain scheduling cannot be used. Examples of such dynamic changes are variations caused by non-measurable variations in the raw material including wear and deterioration of heat conduction caused by coatings in heat exchangers. Adaptive control is the solution to these problems, provided that you have no other problems that will ruin the benefit of adaptive control, such as outside disturbances or valve stiction. A common application for adaptive control is pH control, where adaptive control is used to compensate for the variations in the relationship between the pH value and the controller output. The problem was described in Section 5.6.

There are, of course, cases where the dynamics of the process vary both predictably and unpredictably. All examples of unpredictable dynamic variations above can, for example, be combined with a nonlinear valve in the control loop. For these problems, the best solution is a combination of gain scheduling and adaptive control. Gain scheduling then has to take care of the predictable variations. This means that adaptive control only needs to adapt to the variations that are really unknown.

Adaptive technology is also useful in connection with feedforward. This is described in more detail in Section 6.5.

6 Control Strategies

6.1 Introduction

So far, we have studied individual control loops, that is, a process is controlled by only one feedback loop. Mastering the single loop control by getting the controller, sensor and actuator to work together is a prerequisite for effective process control. However, it is not enough because the individual control loops are related through the process connections.

The single control loops have to interact to achieve the overall process control objective. Connecting individual control loops so that they collaborate and interact well is an advanced engineering art. It is often a knowledge that has been gradually built up by many people over many years of operation and experience. When new production plants are put into operation it is therefore common to copy complete solutions from previous sites. However, the development is never fully completed and instead the process efficiency of the plants has to be improved continuously.

The basis for successful configuration of the control instrumentation in any plant is that you know the process well. Because industrial production processes can look very different this means that there is no single general book on control technology that contains all knowledge or information. However, there are a number of basics control strategies that are often used to let control loops collaborate. In this chapter we will discuss the most common strategies.

6.2 Filters

In the control engineering context, information is sent between sensors, controllers, actuators and other instruments via signals. These signals are in other word 'messengers' and it is important that the signals contain the correct information. Unfortunately, the signals can be distorted in many different ways. Here are some examples:

- Electrical interference adds high frequency noise to the signal.
- Mechanical vibrations spread to sensors and add high frequency noise to the signal.
- Backlash and friction in valves lead to oscillations in control loops. This was described in Chapter 5.
- When converting from an analogue signal to a digital representation, quantisation occurs, that is, the signal is limited by the resolution of the conversion. If this resolution is low, the signal will be distorted.
- Many sensors have built-in low-pass filters. If this filtering is too restrictive, information about sudden changes in the signal will be omitted by the filtered signal.

https://doi.org/10.1515/9783111104959-006

Filtering is used to remove unwanted components in signals and to highlight essential information. There are many different types of filters. Here we will limit ourselves to the most common ones, namely excessive quantisation, low-pass filters, high-pass filters and lead/lag filters.

Excessive Quantisation

A signal that passes through an analogue-to-digital or AD-converter is quantised, which means that it has a limited resolution of its amplitude. In other words, AD-conversion is a type of filtering that changes the properties of the signal. Most often, you want the digital signal to be as good a representation of the analog signal as possible, that is, the quantisation should be as precise and small as possible.

However, there are cases where excessive quantisation can reduce high frequency components in the analog signal. An excessive or granular quantisation means that you introduce a "dead zone", that is, the analog signal must change significantly before the discrete signal is changed to a new value. If the measurement noise is within this dead zone, a large part of the variations caused by the noise will be filtered out. Unfortunately, once you have crossed the border of the dead zone, disturbances become even larger. The effect is illustrated in Figure 6.1.

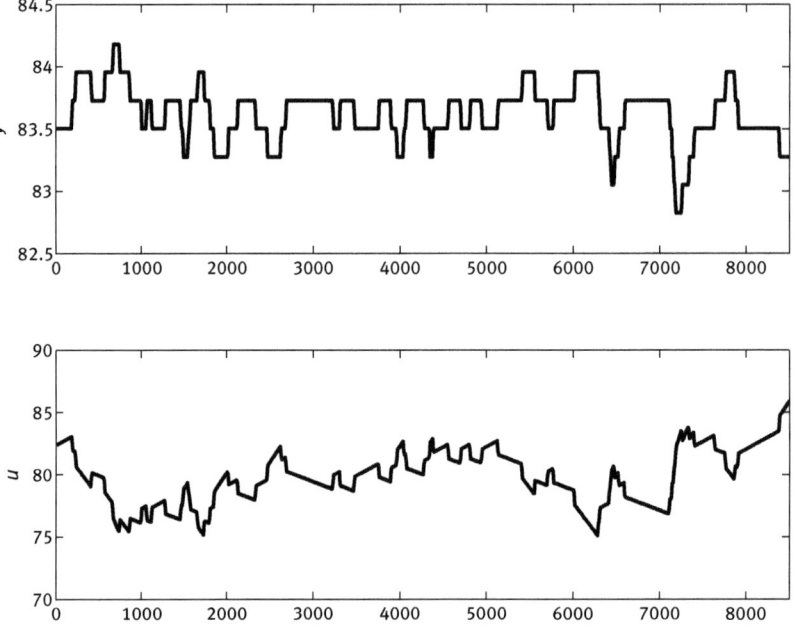

Fig. 6.1: Temperature control with quantisation of 0.2 % in the process variable. Time is in seconds.

The figure shows data from a temperature control loop of an oven where an excessive quantisation of the process variable of 0.2 % was applied. The oven is controlled with a PI-controller with the gain $K = 5$, which causes the controller output to jump by 1 % each time the process variable changes. The disturbances in the process variable and the controller output are probably more significant compared to disturbances that would have occurred without the increase in quantisation. In the vast majority of cases, it is much more efficient to use a low-pass filter to reduce high-frequency disturbances than to increase the quantisation. Low-pass filters are described in the next section.

Low-Pass Filters

A low-pass filter reduces high-frequency noise signals by performing an averaging of the signal. The most common low-pass filter has the same dynamic properties as the first order process described in Chapter 2.

The low-pass filter has a parameter to be specified by the user: the filter time constant T_{lp}. The filter time constant indicates the boundary between the frequencies that are considered low and that are allowed to pass the filter and the frequencies that are considered high and that are not allowed to pass. Roughly speaking, one could also say that the filter time constant indicates for how long the filter should average the signal.

Figure 6.2 shows the step responses for three low-pass filters with different filter time constants T_{lp}. The longer the time constant, the starker or stronger the filtering will be. The choice of T_{lp} depends on how fast the useful part of the signal is that you want to pass through the filter and how fast the disturbance is that you want to filter out. If you select the time constant too short, then the disturbance from the signal is not sufficiently reduced. If, on the other hand, you select the time constant too long, you will filter out useful information. The problem is illustrated in Example 6.1.

Example 6.1.   Low-pass filter

Suppose we have a process variable that consists of two sine signals

$$y = \sin t + 0.5 \sin 50t$$

where the first signal, with a frequency of 1 rad/s and a time period of 6.3 seconds, is the useful component of the signal. The second component has a high frequency, 50 rad/s which corresponds to a time period of 0.13 seconds. The second component is a measurement disturbance that we want to filter out.

Figure 6.3 shows the output signals from low-pass filters with different choices for the filter time constant. The upper left plot shows the unfiltered signal, since the filter time constant is set here to $T_{lp} = 0$ s. The other three plots show the output signal for different choices of filter time constant. The plots show that the longer T_{lp} is, the more the disturbance is reduced.

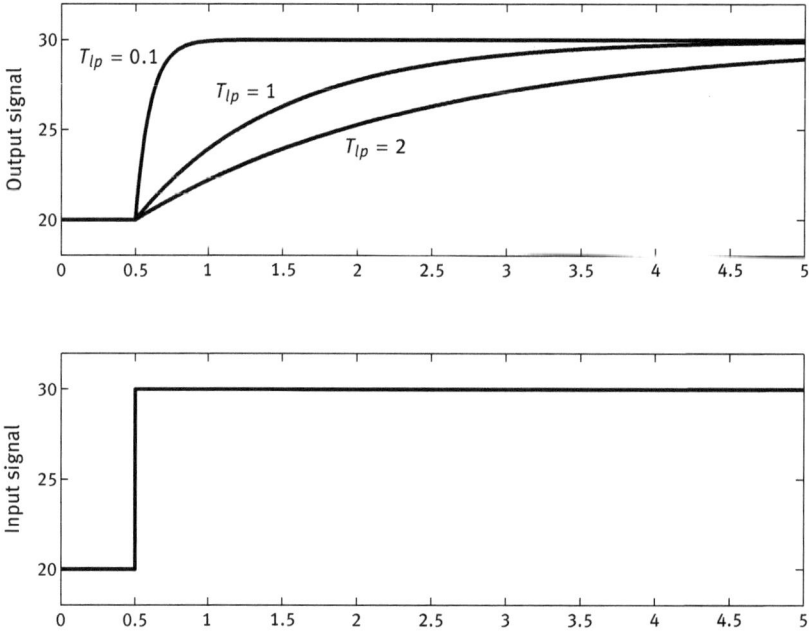

Fig. 6.2: Step response of a low-pass filter with different filter time constants T_{lp}.

The lower right plot in Figure 6.3 illustrates the problem of selecting a filter time constant that is too long. The filter then also affects the useful component in the signal, which here has a lower amplitude and has also been delayed. In this example, the choice $T_{lp} = 0.1\,\text{s}$ for the filter time constant seems to be a good one. □

The choice of filter time constant depends on the frequencies of both the useful signal and the disturbance signal. The choice is easy if these frequencies are very different, but can be difficult if they are close to each other. In such cases, it may be necessary to use more advanced filters than the simple low-pass filter described here. You can, for example, connect several simple filters in series and thus get filters that have a more distinct boundary between the low-frequency signals that are passed through and the high-frequency signals that are filtered out.

High-Pass Filters

The high-pass filter is a filter that lets high frequencies through and filters out low frequencies. Just as for the low-pass filter, a time constant T_{hp} is specified here, which determines the boundary between high and low frequencies. If you choose a large value of T_{hp}, you will let more of the low-frequency signals through, while a short

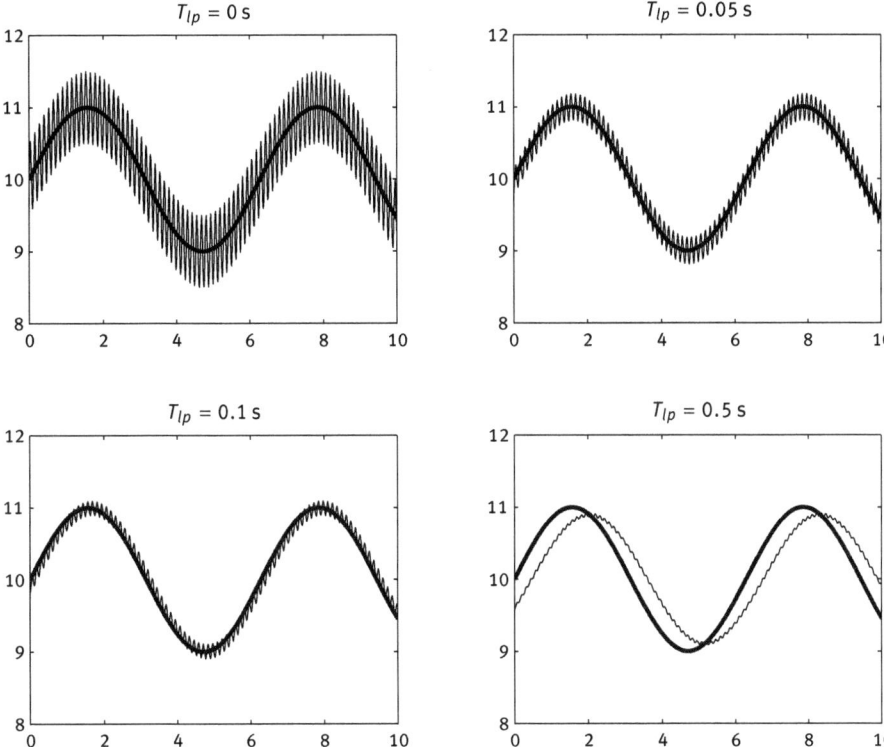

Fig. 6.3: Output signals (thin lines) from low-pass filters with different choices of filter time constant in Example 6.1. The thick lines show the useful component of the signal. The time scale is seconds.

time constant T_{hp} can mean that you filter out some of the high-frequency parts that you really want to let through. This is illustrated in the following example.

Example 6.2. High-pass filter

Suppose we have a process variable that consists of two sine signals

$$y = \sin 50t + 0.5 \sin t$$

These are the same frequencies as in Example 6.1, but now the high frequency signal with the frequency 50 rad/s and the period time 0.13 seconds is the useful component in the signal. We want to filter out the low-frequency component with the frequency 1 rad/s and the time period of 6.3 seconds.

Figure 6.4 shows the outputs from high-pass filters with different choices of filter time constant. The upper left plot shows the unfiltered signal, which corresponds to an infinitely long time constant, $T_{hp} =\rightarrow \infty$. The other three plots show the output signal for different choices of filter time constant. The figures show filters with shorter values of T_{hp} reduce the low frequency disturbance more.

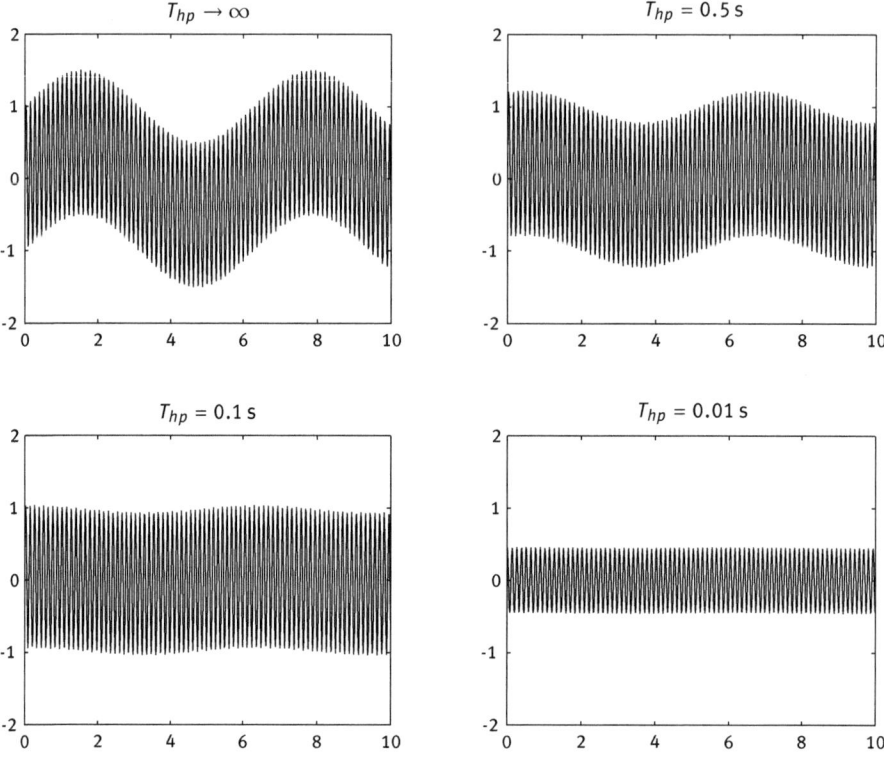

Fig. 6.4: Outputs from high-pass filters with different choices of filter time constant in Example 6.2. Time scale is in seconds.

The lower right plot in Figure 6.4 illustrates the problem when you select a too short filter time constant. The filter also affects the useful component in the signal, which has a lower amplitude in this plot. The choice $T_{hp} = 0.1$ s seems to be a good choice for the filter time constant in this example. □

In the same way as for the low-pass filter, you can get a more efficient filtering if you use more advanced filters. High-pass filters are used in various contexts when you want to get rid of DC voltage levels in signals. High pass filters are, however, an unusual component in process instrumentation.

Lead/lag Filters

The lead/lag filter consists of a series connection of a low-pass filter and a high-pass filter. It therefore has two filter time constants, T_{lp} and T_{hp}, that need to be selected.

Low-pass filtering means that you attenuate high frequencies in the signal at the cost of the filtered signal reacting more slowly to sudden changes in the input signal. This was illustrated in Figure 6.2. This effect can be done in the opposite way with a lead/lag where $T_{hp} > T_{lp}$, that is, the filtered signal can follow sudden changes better at the cost of amplifying high frequencies in the signal. This can be interesting if the signal is relatively free from high frequency disturbances. The following example illustrates this.

Example 6.3. Lead/lag-filtering of a temperature sensor signal
Most sensors have built-in low-pass filters to give the sensor a high level of accuracy. In some cases, the user can choose the filter time constant, but not always. Therefore, the sensor may have a filter time constant that is so long that the sensor reacts too slowly to sudden changes in the process variable. Temperature sensors often have a low noise level. Therefore, you can compensate for an excessive low-pass filtering by using a lead/lag filter. This is illustrated in Figure 6.5.

Figure 6.5 shows a temperature that changes relatively quickly from 10 % to 20 %. The measurement has some uncertainty, which is reflected in a noisy signal. To get a more accurate signal, the sensor has a low-pass filter in the sensor with time constant $T_{lp} = 2\,\mathrm{s}$. The sensor output therefore is a signal given by the dashed curve in the figure. The signal is now free from the high frequency disturbance, but the price is that the change from 10 % to 20 % is much slower than the actual temperature increase.

To obtain a signal that describes the actual temperature more accurately, the sensor signal has been passed through a lead/lag filter. The lead/lag filter parameters are set to $T_{hp} = 2\,\mathrm{s}$ and $T_{lp} = 0.5\,\mathrm{s}$. You compensate for the slow sensor filter because the lead/lag filter is connected in series with the sensor's low-pass filter and you have selected T_{hp} as large as the sensor time constant. The lead/lag filter now introduces a new low-pass filter with the time constant $T_{lp} = 0.5\,\mathrm{s}$ that is four times shorter than the time constant of the sensor's filter. The output of the lead/lag filter therefore

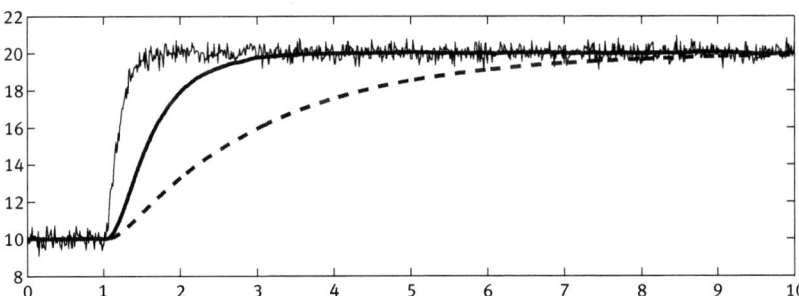

Fig. 6.5: Lead/lag filtering in Example 6.3. The thin line shows the actual temperature. The dashed line shows the low-pass filtered sensor signal and the solid line shows the output signal from the lead/lag filter.

represents the actual temperature better than the sensor signal. This is illustrated in Figure 6.5. ☐

In addition to the above application, lead/lag filters can be used in connection with feedforward control, mid-range control and decoupling, which we will discuss in following sections.

6.3 Selector Control

Selectors are a type of nonlinearity that we introduce in our control systems to handle logical functions. Selectors are sometimes referred to as override control or MIN-MAX-control because the most common types of selectors are the MIN and MAX selectors. These functions select the input signals that have the largest or the smallest value and send this as the output signal. The principle is illustrated in Figure 6.6.

There are also other types of selectors. A selector with three inputs whose function is to select the "middle" signal can be used as protection against sensor failure. If three sensors are connected to this selector and one of them fails, control can continue without interruption. Figure 6.7 shows the principle for a general selector function.

Here, the user can determine the function of the selector by letting the combination of a number of logic inputs l_i determine which of the input signals u_i should be selected. Typical applications for selectors are safety features and operation during

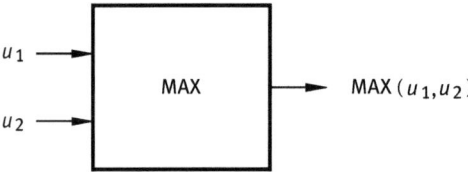

Fig. 6.6: Function of a MAX-selector.

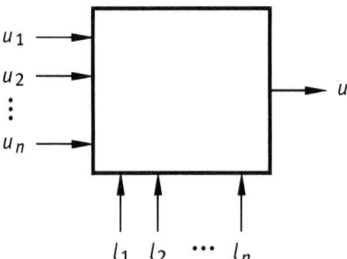

Fig. 6.7: Function of a general selector. Which of the input signals $u_1, u_2, \ldots u_n$ will be selected is determined by the logical input signals $l_1, l_2, \ldots l_n$.

commissioning and shutdowns of facilities. We illustrate the use of selectors with an example.

Example 6.4. Steam pressure control

Figure 6.8 describes the control of the steam pressure from a steam boiler.

The steam pressure is normally controlled by pressure controller PIC2. The set-point of PIC2 indicates the pressure we normally want to keep in the steam pipe. There-fore, the controller changes the position of the steam valve depending on how large the steam consumption is.

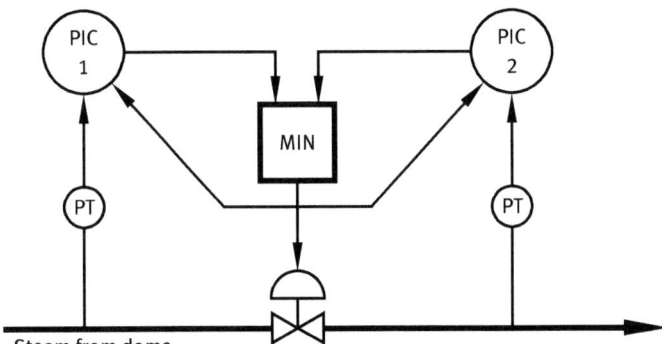

Fig. 6.8: Control of the steam pressure of a steam boiler with a selector controller. PIC, *Pressure Indicating Controller*, is a pressure controller and PT, *Pressure Transmitter* is a pressure transmitter.

At times, it may happen that the steam outlet is so large that controller PIC2 wants to open the valve so much that the pressure in the boiler drops to an unacceptably low level. To solve this problem, you connect another controller, PIC1, according to Figure 6.8. The process variable of this controller is the pressure in the steam pipe from the boiler. The setpoint of PIC1 is the minimum acceptable pressure in the pipe.

By placing a MIN selector as shown in Figure 6.8, the controller now uses the smallest of the two valve positions. Normally the pressure in the boiler is so high that the pressure in the steam pipe is significantly above the minimum acceptable pres-sure. PIC1 then wants to open the valve. PIC2, on the other hand, wants to open the valve not as much as PIC1 and may therefore decide the valve position and thus the pressure in the steam pipe. If, on the other hand, the desired steam outflow should be very large, PIC2 would open the valve widely. At a certain position PIC1 wants a smaller valve opening than PIC2 and will thus take over the pressure control and determine the valve position.   □

Tracking

Selectors can pose the same problem to the controller as actuator constraints, see Section 3.6. Assume, for example, that PIC2 in the above example may suddenly leave the control to PIC1. The controller output from PIC2 will then not reach the valve. This results in the error not being eliminated and if we do not take action, the integral part of PIC2 will grow resulting in integral windup.

As with all other types of nonlinearities, it is important to tell the controller that the nonlinearity exists. This is now not as easy as with constraints, because the nonlinearity depends on a different control loop and therefore changes independently.

The problem is solved by tracking. Tracking means that the controller abandons its previous task of using the controller output to try to get the process variable to be as close to the setpoint as possible. Instead, the controller lets the controller output *follow* another signal.

Tracking is best explained by writing the controller output signal in discrete form, as you would when the controller is a function in a computer. In these cases, we can write the output of the controller as follows:

$$u(t) = u(t-1) + \Delta u(t)$$

The controller output at time t, $u(t)$, is equal to the last issued controller output $u(t-1)$ plus a change $\Delta u(t)$. When the controller is not allowed to control the valve, the controller output instead becomes

$$u(t) = v(t-1) + \Delta u(t)$$

where $v(t-1)$ is the real controller output, that is, the other controller's output signal. In other words, in each new calculation of the output signal, we act as if the output of the selector matched our previous controller output. The controller therefore "tracks" the actual controller output.

Most industrial controllers have the option of connecting a so-called tracking signal to be able to handle tracking. In Example 6.1, the output signal from the MIN selector, which is the real controller output, is fed back to the controller. This is illustrated in Figure 6.8.

6.4 Cascade Control

Cascade control is a control strategy based on the combination of two controllers, where the output signal from one controller is the setpoint of a second controller. Why this may be useful is illustrated in the following example.

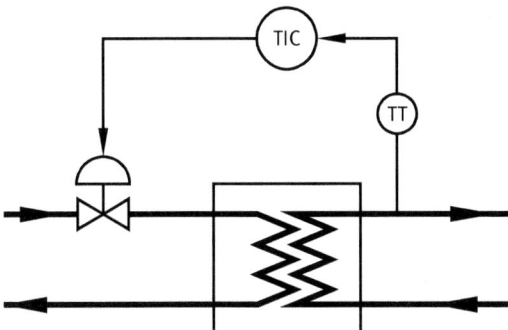

Fig. 6.9: Temperature control of a heat exchanger.

Example 6.5. Control of a heat exchanger

Suppose we have a heat exchanger that sends steam into the primary side to heat water on the secondary side. We want to control the temperature on the secondary side by adjusting the steam valve on the primary side. You can do this by letting a temperature controller operate directly on the steam valve, as shown in Figure 6.9 where the left side is the primary side and the right side the secondary.

What really affects the temperature is not the position of the steam valve, but the steam flow. If the valve is linear and the steam pressure does not vary, this is not a problem, because then there is a linear relationship between the valve position and the steam flow. Usually, however, the valve is nonlinear and the steam pressure varies.

Suppose that, for example, the steam pressure on the primary side suddenly drops. This reduces the steam flow, which means that the water on the secondary side is not heated as much as before. The temperature controller will then give instruction to open the valve wider and after a while the steam flow will be corrected. In other words, this control strategy works but at the price of a large, albeit temporary, deviation.

If you can measure the steam flow, you can connect a flow controller as shown in Figure 6.10. An internal control loop ensures that the steam flow is controlled. The set-point for the flow controller FIC is given by the controller output from the temperature controller TIC. This is called *cascade control*.

Cascade control means that the main controller TIC is given a simpler task. Instead of TIC carrying out all the work, part of the control task is transferred to the flow controller. The TIC controller now only needs to say which flow it wants. Then the flow controller must make sure that we really keep the desired flow. A pressure variation will be smoothed out quickly by the flow controller, which means that the temperature is not disturbed as much as without cascade control. Figure 6.11 shows the simulation of the heat exchanger control.

The dashed lines in Figure 6.11 depict control without cascade coupling according to Figure 6.9. The temperature controller TIC is a PID-controller tuned with the AMIGO

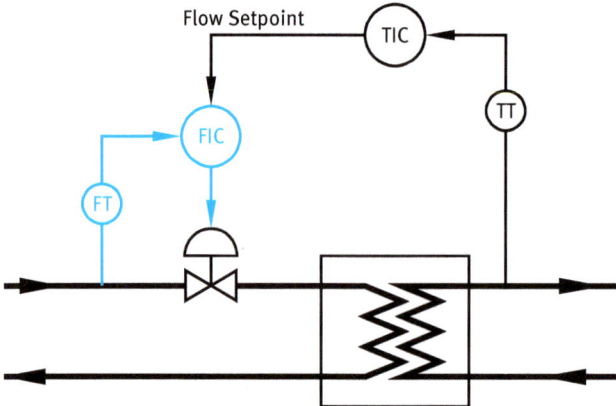

Fig. 6.10: Cascade control of a heat exchanger.

method. The parameters are $K = 1.16$, $T_i = 10.2$ and $T_d = 2.58$. The simulation clearly shows how the pressure drop causes a decrease in flow and thus a disturbance in the temperature.

The solid lines in Figure 6.11 illustrate the benefit of cascade control. The flow controller FIC is a PI-controller and the AMIGO method gives the parameters $K = 1.67$ and $T_i = 1.34$. Because the flow controller steers the flow quickly back to its correct value after the pressure disturbance, the disturbance in the temperature becomes very small.   □

The general principle of cascade control is shown in Figure 6.12. The primary objective is to control process variable y_1 using controller C_1. This could have been done by using only controller C_1 and let its controller output pass directly to the process. In cascade control, you take advantage of having access to an additional process variable, y_2. By providing a local feedback from y_2 via the controller C_2, you can achieve a much more efficient control than with a single PID-controller. Controller C_1 is often called the primary controller and controller C_2 the secondary controller. Other designations are master controller and slave controller.

Cascade control is a typical example of how the simple structure of the PID-controller can achieve more advanced control solutions by combining several controllers and measurements. The main reason for using cascade control is that in this way you can deal faster with disturbances that come into process section P_2, before they have time to cause disturbances in the primary process variable y_1. An example of this is the pressure variations in Example 6.5. Of course, a prerequisite for this to work is that the internal or secondary control loop is significantly faster than the external or primary control loop.

Another advantage of cascade control is that the dynamics of the process controlled by the primary controller can be simplified. Without cascade control, the con-

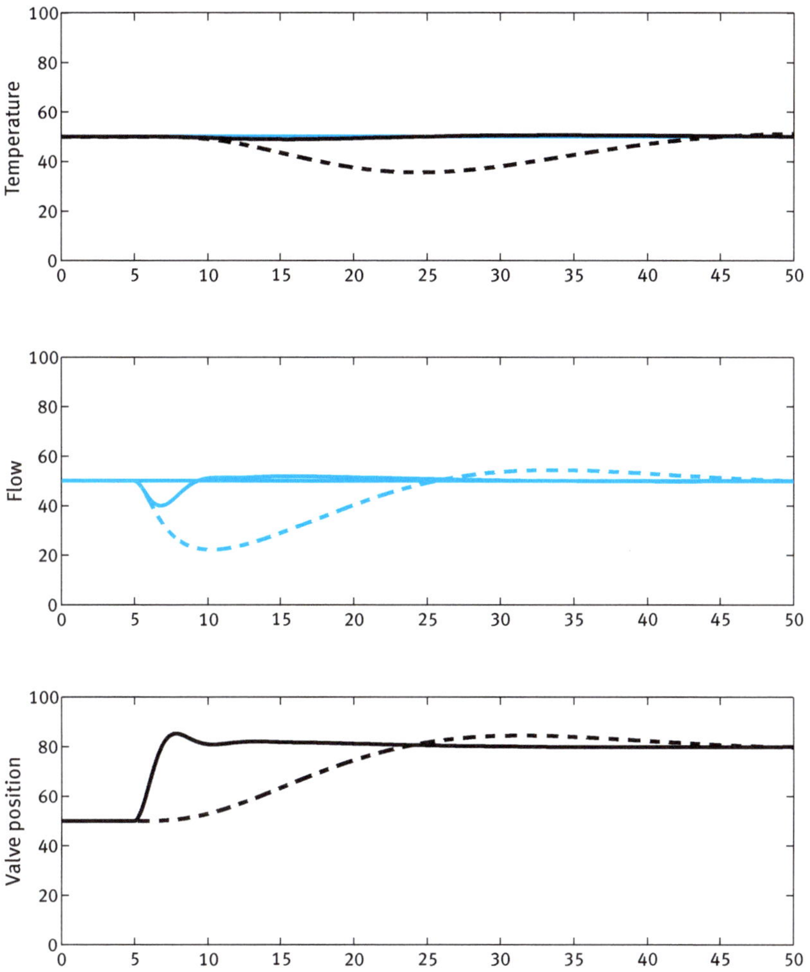

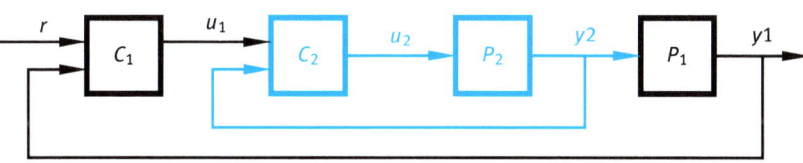

Fig. 6.11: Control of a heat exchanger in Example 6.5 with cascade control (solid lines) and without cascade control (dashed lines). The figure shows the control of a load disturbance at time $t = 5$.

Fig. 6.12: Principle of cascade control.

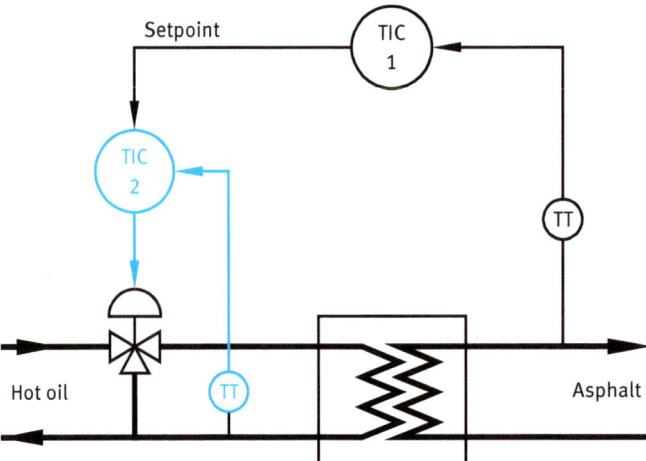

Fig. 6.13: Cascade control of a heat exchanger.

troller C_1 works against a process consisting of the process sections P_1 and P_2. With cascade control, the process section changes to the combination of P_1 and P_2 fed back with C_2. In many cases, this dynamic is simpler.

We will now study another example of cascade control of a heat exchanger.

Example 6.6. Control of heat exchangers

Figure 6.13 shows another example of cascade control of a heat exchanger. In this case, asphalt is heated on the secondary side using hot oil on the primary side. Here, you do not have the pressure and flow variations on the primary side as in the previous example because of the liquid nature of the heating medium. However, there is another serious load disturbance, namely the temperature of the incoming asphalt. These disturbances would not be detected by the secondary controller with the cascade coupling illustrated in Figure 6.10.

On the other hand, if you measure the temperature of the hot oil that leaves the heat exchanger as shown in Figure 6.13, one can catch a disturbance in the asphalt temperature early on. For example, suppose the temperature of the incoming asphalt suddenly drops. Then the hot oil cools off more than before. By connecting the temperature of the exiting hot oil to a secondary controller in a cascade loop, you will quickly compensate for the cooler asphalt by turning the three-way valve and increasing the oil flow into the heat exchanger. □

Another common area of the use of cascade control are nonlinear valves. Non-linear valves can be compensated for via gain scheduling or any of the other methods described in Chapter 5. If there is no possibility of applying any of these methods, you can use cascade control instead. The solution requires that we have a flow meter that reads out the actual flow through the valve. The secondary controller takes this flow measurement as its process variable and controls the valve. If the valve is nonlinear, you can tune the controller according to the worst case dynamics. In this way, the nonlinearity does not have to be taken into account by the primary controller.

Another application of cascade control is valve positioners. A valve positioner is nothing more than a controller that is located as a secondary controller in a cascade configuration where the primary control loop controls the process variable. The valve positioner is the secondary controller, often a P-controller, whose process variable is the valve stem position.

Tuning of Cascade Controllers

When tuning two controllers that are in cascade mode it is important to tune them in the correct order as follows:
1. First, put the primary controller C_1 into manual control.
2. Second, tune the secondary controller C_2.
3. Third, put the secondary controller C_2 into automatic mode.
4. Finally, tune the primary controller C_1.

The process section controlled by controller C_1 consists of section P_1 plus section P_2 fed back with controller C_2. This is shown in Figure 6.12. Since C_2 is part of the process that C_1 controls, we must tune C_2 before C_1. While we are tuning C_2, it is advantageous to put C_1 in manual control, so that the control of C_1 does not interfere with the experiments.

Also to keep in mind when doing cascade control is that the two loops usually have different types of disturbances. The inner loop is normally disturbed by changes in the setpoint, that is, by the controller output from the external controller. When tuning this controller, you should therefore make sure that you do not get too large overshoots in the event of fast setpoint changes. This can be done, for example, by selecting an appropriate setpoint weight. The outer loop is often more affected by load disturbances.

It often suffices to use a P-controller as a secondary controller. It mostly does not matter that you get a persistent control error in the secondary loop due to the lack of integral action. However, if the control error becomes large because the gain in the secondary loop is low, you can get into difficulty. It is not guaranteed that the entire control range of the primary controller output can be used.

Industrial control vendors usually have a setting called external and internal control. External control means that the setpoint is determined by an outside algorithm while internal refers to a single loop where the setpoint is specified by the user. The inner loop in cascade control is always in external mode. External control, however, does not necessarily imply cascade control. The external setpoint could be calculated from a different advanced strategy such as model predictive control.

Commissioning

Commissioning a cascade controller must also take place according to a special pattern if you want to avoid sudden changes in the process variable. Assume that both controllers are in manual control and that we want them in automatic control. We will then work "from the inside out" in the following way:

1. Set the setpoint of C_2 to its current process value.
2. Select internal setpoint in C_2 and put C_2 in automatic control.
3. Change the controller output of C_1 so that it matches the setpoint of C_2.
4. Change C_2 from internal to external setpoint.
5. Set the setpoint in C_1 to its current process value.
6. Put C_1 to automatic control mode.
7. Adjust the setpoint of C_1 to the desired level.

Industrial control systems that have a cascade control built into them automatically handle some of the above transitions.

Tracking

In cascade control, we have special problems that arise in the external controller when the internal controller does not work as intended. There are essentially three cases that need to be taken care of:

1. If the internal controller's output is at its constraint or limit.
2. If the internal controller switches from following the external setpoint from the outer controller to following its own internal setpoint.
3. If the internal controller is switched from automatic control to manual control.

In all these cases, the cascade connection no longer works and the output signal from the primary controller can not influence the control error. If the primary controller does not know about it, the result can be integral windup.

In modern control systems, these problems are solved automatically, but if you have older systems or single-loop controllers with which you build up the cascade control yourself via two independent controllers, you have to solve the problems yourself. This requires the ability of the primary controller to receive a tracking signal.

If the output signal of the secondary controller is constrained, the process variable of the secondary controller should be selected as the tracking signal in the primary controller. There is also a need for a digital information from the secondary controller to the primary controller that tells when tracking occurs. At these times, the primary controller will work around the secondary controller's process variable instead of around its own previous controller output.

When switching the secondary controller to follow its local setpoint, the local set-point should be sent back as a tracking signal to the primary controller instead of the external one. In this way, both integral windup and bumps when returning to cascade control are avoided.

When the secondary controller switches to manual control, the process variable of the secondary controller should be sent back to the primary controller as a tracking signal.

Another easier way to solve the last two cases is to simply let the primary controller switch to manual control at these times. In this way, the integral part of the primary controller is prevented from growing. However, you are not protected against bumps when returning to cascade control. This solution requires a digital signal transmission. The internal controller can usually issue a digital signal that indicates that an internal setpoint is used or that the controller is in manual control. The external controller can be switched from automatic control to manual control via a digital input signal. By connecting these two signals, we can put the external controller in manual control when the internal controller is put in manual control or has an internal setpoint.

The reason why we can use this simpler method in the last two cases is that we have voluntarily chosen to break the cascade connection. It then does not matter that the primary controller is in manual control. If, on the other hand, we would put the primary controller in manual control when the secondary controller's controller output is constrained, we will no longer react to changes in the primary controller's setpoint, but "get stuck" in manual control.

6.5 Feedforward

Feedback is an effective method for solving many control problems: You measure the signal to be controlled, compare it to a setpoint and then calculate a controller output based on this comparison. However, a limitation of the feedback principle is that the controller does not react to a disturbance until a control error has already occurred. In many cases, it is possible to measure a disturbance that will affect the control loop. By compensating for the disturbance even *before* a control error has occurred, you can often significantly improve the control performance. This technique is called feedforward.

A well-known example of the use of feedforward is the control of the temperature in residential buildings. Most houses have a temperature sensor that measures the outdoor temperature. The outdoor temperature is the largest and most important disturbance when controlling the indoor temperature. By using the information about the outdoor temperature, you compensate for variations in this temperature even before it has affected the indoor temperature. If, for example, the outdoor temperature drops, this should result in an increase in the temperature in the radiators, even though the indoor temperature has not yet started to drop. With the help of feedforward, the variations in the indoor temperature can be significantly reduced.

How early you can measure the disturbance is crucial when determining how effective feedforward will be. If the disturbance can be measured long before its effect is seen in the process variable, feedforward is effective, especially if the process has a long dead time. If you measure the disturbance so late that its effect is already visible in the process variable, there is usually no point in using feedforward. In that case, it is just as well to let the controller handle the disturbance via feedback. We will illustrate this in more detail in the next example.

Example 6.7. Concentration control

Figure 6.14 shows a concentration control loop, where an acidic and an alkaline (base) flow are mixed. The concentration is controlled by the alkaline flow which is adjusted in by a concentration controller.

The control problem can be solved in the usual way using feedback only. For example, if the acid concentration increases, this will give rise to an incorrect concentration and a control error. The controller will therefore adjust the alkaline flow so that the correct concentration is again achieved. However, this happens at the cost of the concentration being disturbed and deviating from the setpoint for a period of time. Through feedforward, however, we can increase the alkaline flow as soon as we notice a change in the acid concentration, hopefully well in time before a control error has time to occur.

Assume that the dead time plus the dominant time constant between the valve and the mixture concentration sensor is T_1, that is, it takes about T_1 seconds before a change in the valve position gives rise to a reaction in the concentration sensor. Assume further that the dead time plus the dominant time constant between the acid concentration sensor and the mixture concentration sensor is T_2. The efficiency of feedforward control depends on the relationship between the two times T_1 and T_2.

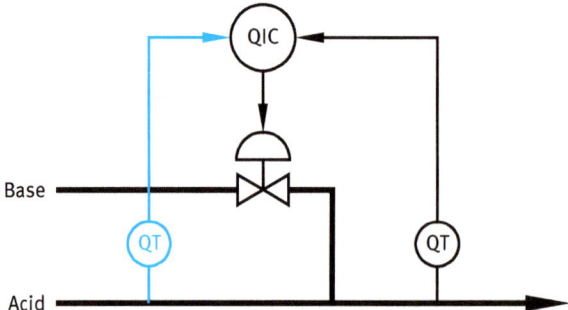

Fig. 6.14: Feedforward concentration control. QIC, *Quality Indicating Controller*, is a concentration controller and QT, *Quality Transmitter*, is a concentration sensor.

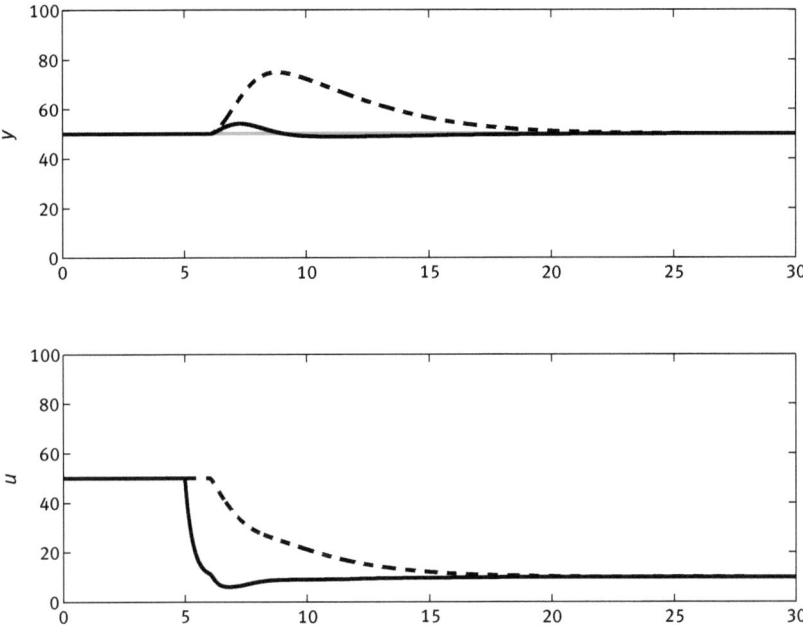

Fig. 6.15: Concentration control of Example 6.7. The figure shows control with feedforward (solid lines) and without feedforward (dashed lines) when a load disturbance occurs at time $t = 5$.

$T_1 \gg T_2$ A change in the acid concentration will be seen in the mixture concentration sensor a relatively short time after we have detected it in the acid concentration sensor. In this case, the improvement with feedforward is small. The controller can solve the control problem equally well via feedback only. It may be justified to use feedforward if for some reason we have been forced to choose a slow controller tuning, but still want a quick control of this disturbance.

$T_1 \approx T_2$ In this case, the feedforward is effective. Through the feedforward control the controller can prevent the mixture concentration from changing significantly by immediately making a compensation when the acid concentration changes.

$T_1 \ll T_2$ If we were to make a compensation immediately when we detect a change in the acid concentration, this would mean that the alkaline flow changes too early and therefore results in an incorrect concentration of the mixture. In this case, we must therefore delay the compensation. The easiest way to do this is to delay the feedforward signal (the acid concentration signal) by time $T_2 - T_1$.

Figure 6.15 shows a simulation of the concentration control loop. The controller is a PID-controller tuned with the AMIGO method. The controller parameters are $K = 0.566$, $T_i = 2.22$ and $T_d = 0.835$. The dashed curves show control without feedforward in the event of a sudden decrease in the input acid concentration at time $t = 5$. The disturbance causes the alkaline concentration in the mixture to rise. The controller compensates for this by restricting the alkaline flow. The concentration is corrected after roughly fifteen seconds.

The solid curves show control with feedforward. Because of feedforward control, the controller can restrict the alkaline flow *before any disturbance has been noticed in the mixing sensor. This means that the disturbance of the mixture concentration is considerably smaller than without feedforward. The example shows a case where $T_1 \approx T_2$, that is, a case where feedforward is effective.* ☐

The above example shows that feedforward pays off if you can measure the disturbance early enough.

Example 6.8. Three-element controller

A classical example of feedforward is three-element control. Figure 6.16 shows the principle of level control in a steam boiler. A valve controls the feed water flow into the boiler dome and another valve controls the steam flow out of the boiler. The control problem consists of controlling the feed water valve so that the level in the boiler is kept constant. The steam outflow varies according to demand and can therefore be seen as a disturbance in the control loop.

The simplest way to control the water level would be to use only a level sensor and use the signal from this to control the feed water valve via a level controller. This is called one-element control. However, level control of a boiler is a technically difficult

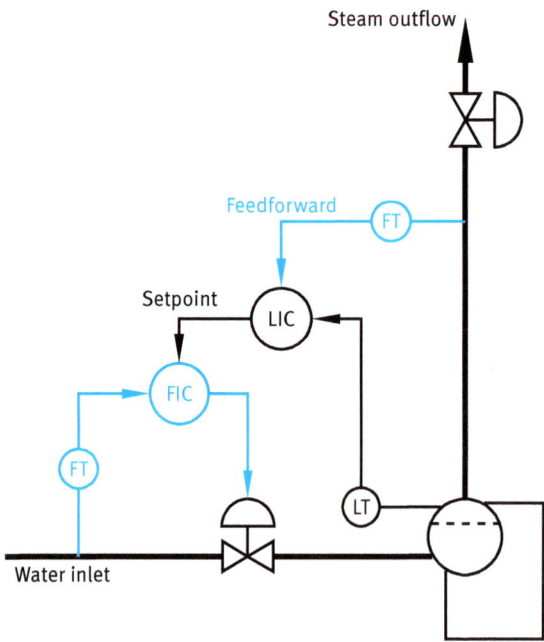

Fig. 6.16: Three-element control.

problem. The process has a reverse step response, see Section 2.3, which means that it has a step response that initially goes in the "wrong direction". An increase in the feed water flow means that the dome level first drops due to cooling effects, and then increases after a while. In addition, an increase in the steam output initially results in an increase in the water level due to the reduced pressure in the boiler. The water level drops only after some time has elapsed due to the increased steam output. Clearly, these reverse responses present problems for the controller.

Due to the difficult dynamics of the process one normally does not use a single loop control strategy to maintain the level. Instead, you measure the steam flow and the feed water flow in addition to the level ("three elements"). The feed water flow is controlled with a cascade connection to the level according to Figure 6.16. The steam flow is fed forward to the level controller. If, for example, the steam outlet should increase, the feedforward signal causes the setpoint of the flow controller to increase. Thus, the feed water valve will open relatively quickly. This feedforward action counteracts the awkward dynamics of the process. This means that the controller wants to close the feed water valve via feedback when the steam flow increases. □

Example 6.9. Heating control

Figure 6.17 shows a simplified process schematic of a heating control in a district heating plant. The control objective is to keep the temperature of the outgoing district heating water constant despite variations in the load of the network. The load variations that occur during normal operation often place great demands on the power control. The boiler's power control consists of a pressure controller that controls the fuel/air supply. The temperature of the outgoing district heating water is controlled via a shunt valve on the primary side of the heat exchanger.

The problem with this power control strategy is that the control system becomes sluggish and does not respond well to load disturbances. Assume that the heat de-

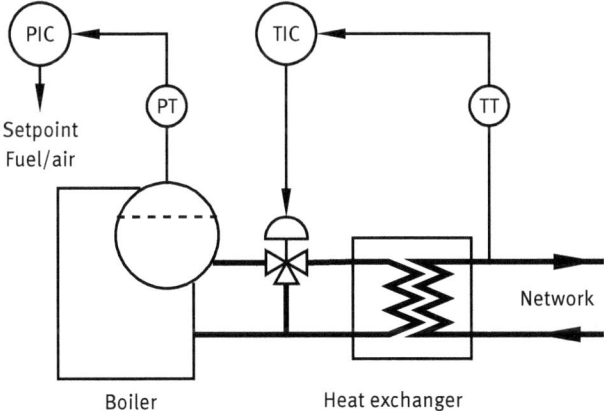

Fig. 6.17: Example of a boiler control.

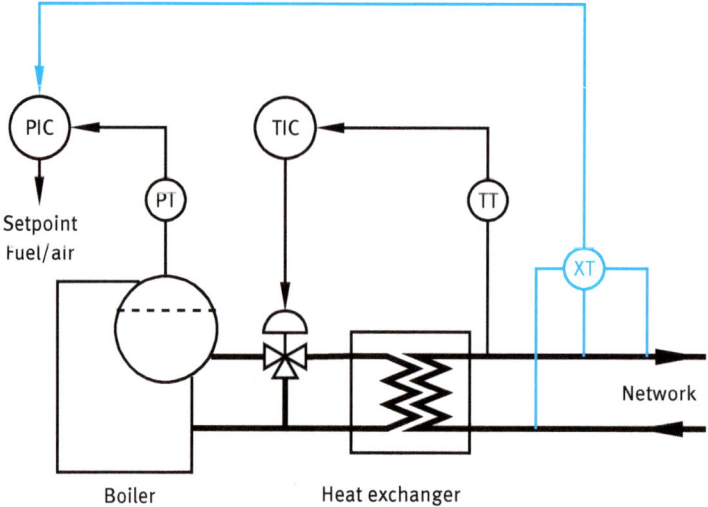

Fig. 6.18: Boiler control with feedforward of a power signal. XT is a power sensor, where the power is determined from the flow and the difference between the temperatures in the inlet and outlet flows.

mand suddenly decreases, which results in the water returning from the network suddenly becoming warmer. The temperature controller will then throttle the valve to bring the water temperature down again. Because this means a smaller heat withdrawal from the boiler, the boiler pressure will now rise. Due to the rising boiler pressure, the boiler pressure controller finally lowers the fuel and air supply. In other words, this control strategy works, but it takes a long time before the essential control objective is achieved, namely to compensate for the reduced heat consumption with a corresponding reduction in the power supply.

The problem can be solved by using the power signal from a power meter that measures the power demand on the district heating network after the heat exchanger. This is illustrated in Figure 6.18. By adding the power signal to the setpoint of the fuel/air controller using feedforward, a direct compensation for load changes is obtained when these occur. This more robust power control has shown to handle significantly larger load disturbances while the variation in the district heating temperature can be kept within reasonable limits. □

General Feedforward Control Structure

Figure 6.19 shows the block diagram of the principle of feedforward. We start from the single control loop, where a measurable disturbance v affects the process. The objective is to take advantage of the fact that the disturbance can be measured to improve the control of the process. Ideally, the controller output should act so that the effect of

v never appears in the process variable y. A realisation of this ideal compensation is in most cases impossible. The compensation shown in Figure 6.19 is usually satisfactory. The measured disturbance is multiplied by a gain of K_f so that the forward signal is computed as

$$u_f = K_f v$$

This signal is added to the controller output signal of the feedback controller. Sometimes the disturbance v should be delayed, so that the compensation does not take place too soon. This is the case when the time it takes for the disturbance to affect the process variable is longer than the sum of the process dead time and the dominant time constant. In this case, if we do not delay v the feed of the feedforward will affect the process variable too early; before the disturbance has impacted on the process.

An appropriate value of the feedforward gain K_f can be determined as follows. Note u when the control loop is in steady-state for two different disturbance levels v_1 and v_2. In the case of concentration control, the output of the controller is thus determined for two different values of the disturbance concentration. By examining the effect of the change in the disturbance level on the controller outputs u_1 and u_2, we can then obtain an estimate of a suitable feedforward gain K_f:

$$K_f = \frac{u_1 - u_2}{v_1 - v_2}$$

This choice of feedforward gain means that we will momentarily adjust the controller output to the expected new level when the disturbance changes. It can be sometimes wise to choose it a little lower.

Note that the gain sometimes becomes negative. The feedforward gain is selected here based on the steady-state conditions of the control loop. In most cases, this provides significantly improved control and is fully adequate. However, you can get an even better control if you take into account the dynamics of the process and the controller and there are tuning rules for feedforward that do this. If the static feedforward is not efficient enough, you can supplement the gain K_f with a lead/lag filter. These filters were described in Section 6.2.

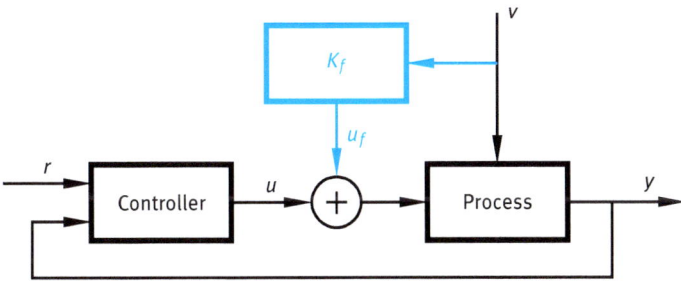

Fig. 6.19: Principle of feedforward control.

Practical Realisation

The principle of feedforward is described in Figure 6.19. This is a good illustration to show the principle, but it is not a good representation for practically implementing the feedforward function. That this is not the case can easily be demonstrated by considering a simple example.

Assume that the process is controlled by a valve that is fully open when the input signal is 100 %. This input signal is given according to Figure 6.19 as

$$u + u_f = u + K_f v$$

This means that the controller output u is constrained when it reaches the value

$$u = 100\% - K_f v$$

Assume for simplicity that $K_f = 1$. If the disturbance at a certain time has the value $v = 70\%$, then the valve is fully open already when the controller output is $u = 30\%$. If the controller does not know this, there is a risk of integral windup. There are also other disadvantages of having a design where the controller output u does not correspond to the signal that actually enters the process. For example, if the controller is in manual control, you want the controller output to determine the position of the valve. This is not the case if you have the structure shown in Figure 6.19.

Feedforward should instead be realised as shown in Figure 6.20. Instead of multiplying the disturbance v by the gain K_f and adding this signal to the controller output u, you should multiply the *changes* in v by K_f and then add these changes to *changes* in the controller output u.

In Section 6.3, we described the controller output as follows:

$$u(t) = u(t - 1) + \Delta u(t)$$

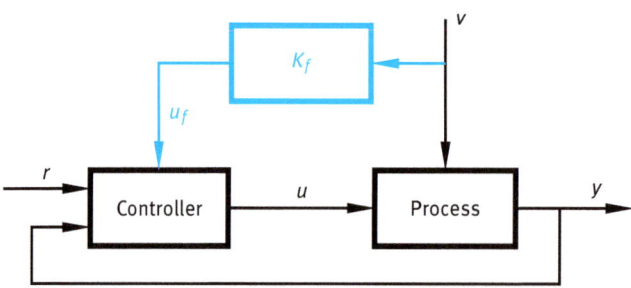

Fig. 6.20: Practical realisation of feedforward control.

The controller output at time t, $u(t)$, is equal to the last issued controller output $u(t-1)$ added with an addition $\Delta u(t)$. When we do feedforward, this control law should be supplemented with a forwarding term:

$$u(t) = u(t - 1) + \Delta u(t) + K_f \Delta v(t)$$

To achieve this, one cannot add the feedforward signal to the controller, but the feedforward signal must be handled inside the controller block as shown in Figure 6.20.

Unfortunately, many system vendors have chosen to implement the feedforward function according to Figure 6.19 instead of according to Figure 6.20. In some cases, an attempt is made to reduce the problem by high-pass filtering the feedforward signal, but this only gives a small improvement of the basically poor solution. Feedforward is a very powerful function that can provide major improvements in control in the process industry. However, the fact that it is often not implemented correctly has unfortunately limited its use.

Feedforward with Cascade Control

When using feedforward in connection with cascade control, it is important that the disturbance is fed to the correct controller. The problem is illustrated in Figure 6.21. A disturbance that enters the process section P_1 must be fed to the primary controller C_1. One might think that it would be more efficient to provide the disturbance to the secondary controller C_2 as this would provide a faster compensation. It is the controller C_2 that is closest to the process and whose output signal directly affects the actuator.

Assume that the fault v_1 is fed to the controller C_2. A change in v_1 then results in the controller C_2 compensating for this and the controller output u_2 will change. So far, so good. However, the next thing that happens is that the controller C_2 detects that

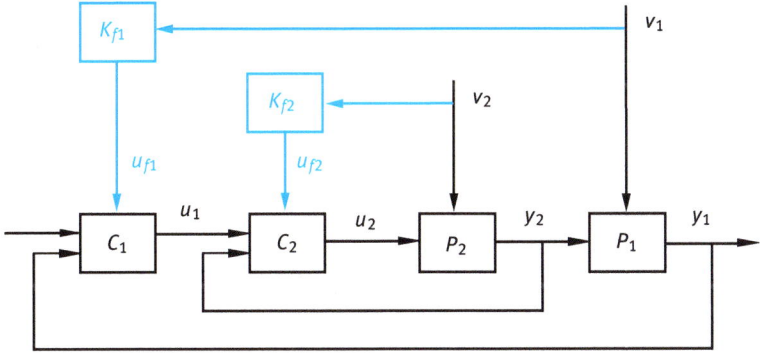

Fig. 6.21: Feedforward for cascade control. Disturbances entering process section P_1 must be fed to controller C_1 and disturbances entering process section P_2 must be fed to controller C_2.

y_2 deviates from its setpoint. Therefore, the controller will undo its control action and quickly change u_2 back to its original value again. The problem is that the controller C_2 has a setpoint determined by the primary controller C_1 which with this solution does not change when the disturbance changes. Therefore, the feedforward connection must be made to the primary controller C_1 so that the setpoint of the secondary controller is changed. In this way, C_2 will adjust y_2 to its new setpoint and you will get a permanent change that will compensate for the disturbance.

Disturbance v_2 in Figure 6.21, on the other hand, which enters the process section P_2, can be fed forwarded to the secondary controller C_2. In most cases, feedforward to the secondary controllers does not result in a big improvement, since the secondary loops in the cascade connection should be fast in relation to the primary loops. This means that the secondary controller compensates for disturbances so quickly that they do not have time to give any major effect in the process variable of the primary loop.

Adaptive Feedforward

Adaptive technology can also be used to handle the feedforward part of the controller. It is often difficult to tune the feedforward parameters manually. In order to be able to do this, you need to find out what effect the disturbance has on the control performance. What causes problems is that you often cannot influence the disturbance. When tuning the PID-controller you can examine the dynamics by making setpoint changes or introducing load disturbances in the control loop. In the feedforward case, you often have to wait for disturbances before you can determine if the parameters are appropriately selected. For the same reason, automatic tuning of the feedforward parameters cannot be made at a point in time chosen by the operator.

Adaptive feedforward, on the other hand, is ideal for tuning the parameters in the feedforward part, even in the case when the dynamics are constant. The adaptive algorithm is constantly waiting for disturbances and adjusts the parameters when the disturbances appear. The principle of adaptive feedforward control is shown in Figure 5.14, with the difference that not only controller output and process variable but also the disturbance is used to determine the process model.

6.6 Mid-Range Control

Mid-range control is a control strategy used when there is more than one actuator to control a process variable. For the sake of simplicity, we will limit ourselves here to the most common case with two actuators and one process variable. Mid-range control is used to determine how the two actuators can be synchronised.

Example 1   Example 2

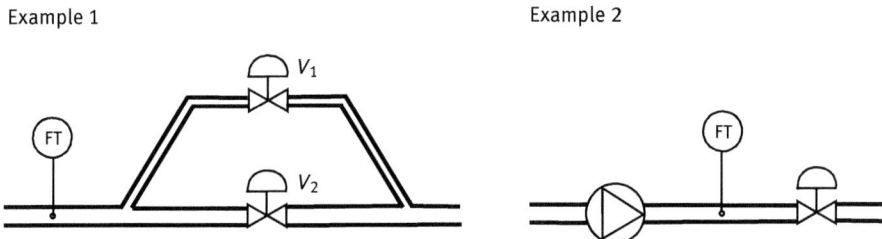

Fig. 6.22: Two common control cases that are solved with mid-range control.

Figure 6.22 shows the two most common problems handled with mid-range control. Example 1 is a case where you want to control a large flow range with high precision. Because the flow can be large, a large valve or pump is required. However, this valve or pump cannot handle the required high precision. To solve the problem, a part of the flow is diverted into a smaller pipe with a smaller and more accurate valve. Assume that the small valve V_1 is in the middle of its operating region and that there are only small disturbances to the flow. In this case, you can use a single controller that controls V_1 to solve the control problem. But if the disturbances are large, then valve V_1 may be constrained and cannot compensate for the disturbance. In this case, the large valve V_2 must also be used for the control and adjust the flow so that the small valve leaves its constraint. Mid-range control strategy is also sometimes called "big-valve-little-valve".

Example 2 in Figure 6.22 shows a case where a flow is controlled with both a pump and a valve. To save energy, you want the valve to be as close to fully open as possible, so that the pump does not have to work at an unnecessarily high pressure.

Valve Position Controller — VPC

The most common way to solve mid-range problems is to introduce a valve position controller (VPC). The solutions for the two examples in Figure 6.22 are shown in Figure 6.23. In Example 1, the flow controller FIC controls the flow to its desired setpoint r_y using the small valve V_1. You want the controller output from the flow controller to be in the middle of its operating region so that it is prepared to act on disturbances and setpoint changes. This is handled by introducing a valve position controller, VPC. The process variable of this second controller is the controller output from the flow controller. The controller tries to steer its process variable to setpoint r_u by adjusting the large valve V_2. If both controllers have integral action, the flow will be steered to its setpoint r_y and valve V_1 will be at its desired value r_u when the loops are in steady-state.

Note that the difference between a VPC and a valve positioner, described in Section 5.3, is that the valve positioner is a controller connected to the valve which controls the valve stem position directly so that it corresponds to the controller output. A VPC, on the other hand, is an external controller that controls the controller output, not the stem position.

In Example 2, the flow is kept at its setpoint r_y by a flow controller that controls the valve. Here, we want the valve to be almost fully open when the control loop is in steady-state. We achieve this in the same way as in Example 1 by letting a valve position controller control the pump to steer the flow controller output to its setpoint r_u. This setpoint is often selected so that the valve is approximately 90 % open.

Figure 6.24 shows the block diagram describing the function of the mid-range control with a valve position controller. The process P_1 and the controller C_1 form a fast control loop. The valve position controller C_2 controls the controller output u_1 of controller C_1, and thus the position of the small valve, by influencing the flow y. This means that the output signal from C_1 is controlled by temporarily driving the process

Example 1   Example 2

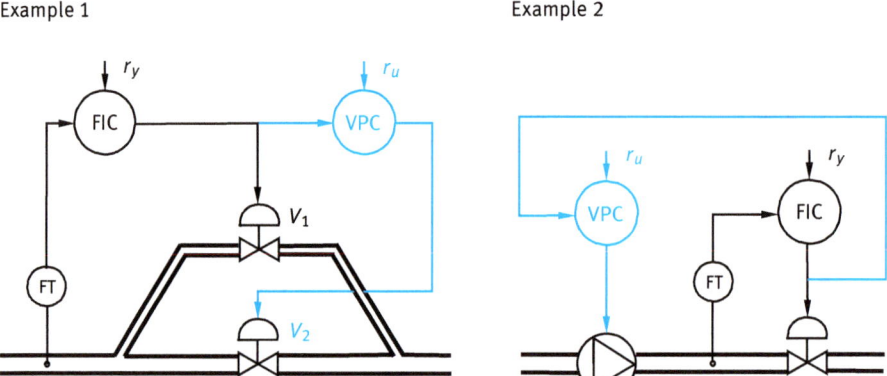

Fig. 6.23: Mid-range control using a valve position controller (VPC).

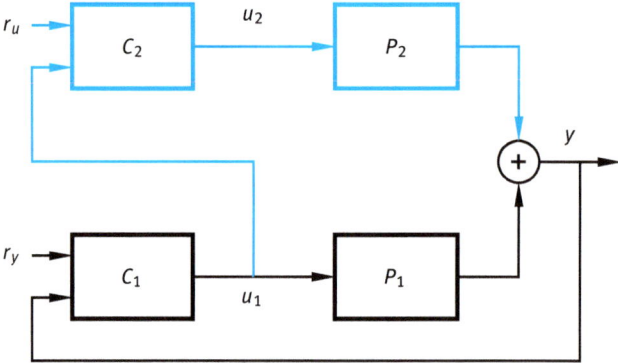

Fig. 6.24: Block diagram of mid-range control with a valve position controller.

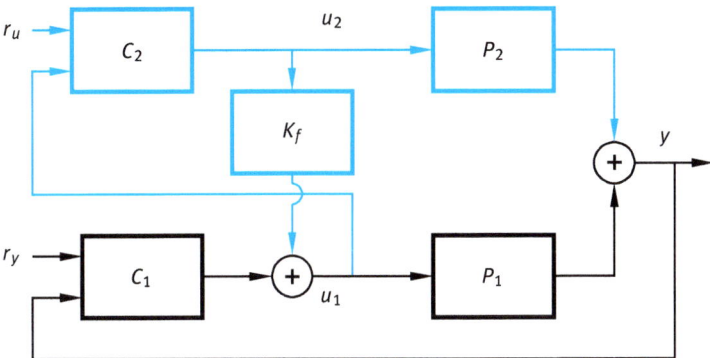

Fig. 6.25: Mid-range control with a valve position controller and feedforward.

variable y away from its setpoint r_y. If this is done slowly, the deviation remains small. One way to reduce the deviation is to use feedforward as shown in Figure 6.25.

In Figure 6.25 a feedforward connection is introduced from the controller output u_2 to the controller C_1. Through this connection the fast controller C_1 finds out what disturbances the controller C_2 introduces and can compensate for them. An appropriate value for the feedforward gain is

$$K_f = -\frac{K_{p2}}{K_{p1}}$$

where K_{p1} and K_{p2} are the process gains in the process sections P_1 and P_2. This steady-state feedforward can be improved by introducing an additional lead/lag filter. This filter was described in Section 6.2. If this is done, it is recommended to set the high-pass filter time constant to $T_{hp} = L_1 + T_1$ where L_1 and T_1 are the dead time and the time constant for the process P_1. The low-pass filter of the lead/lag filter should have a time constant of $T_{lp} = L_2 + T_2$ where L_2 and T_2 are the dead time and the time constant of the process P_2.

The controller C_1 can be tuned in the usual way, for example with one of the methods described in Chapter 4. The controller C_2, however, should be tuned conservatively so that the movements of the large valve do not disturb the process variable too much. It is further recommended to set the controller gain to a low value and let the integral part perform most of the control task. It may also be advisable to add a dead zone to controller C_2, so that the controller stops changing its output signal when the measured value u_1 is close enough to its setpoint r_u. This way, you avoid problems with friction and backlash in the valve.

Example 6.10. Mid-range control with a valve position controller

Figure 6.26 shows a simulation of the problem in Example 1 in Figure 6.22 using the mid-range strategy described in Figure 6.25. The process P_1, which contains the small, fast valve, has a process gain $K_{p1} = 0.2$ and consists of two first order processes in

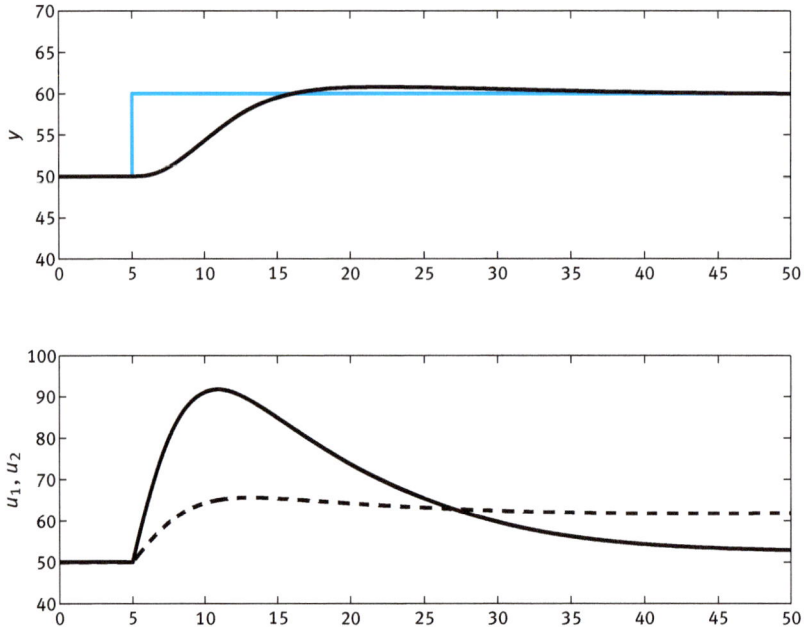

Fig. 6.26: Mid-range control with valve position controller in Example 6.10. The upper figure shows the flow and its setpoint. The setpoint changes at time $t = 5$ s from 50 % to 60 %. The lower figure shows the controller outputs u_1 (solid line) and u_2 (dashed line).

series with time constants of two seconds. The process P_2, which contains the large valve, has a process gain $K_{p2} = 0.8$ and consists of two first order processes in series with time constants of ten seconds.

The controller C_1 is a PI-controller tuned with the AMIGO method. The controller parameters are $K = 8.35$ and $T_i = 2.68$ s. The controller C_2 is also a PI-controller. It is tuned to provide gentle control with low gain. The controller parameters are $K = 0.3$ and $T_i = 20$ s. Both controllers have the setpoint weight $b = 0$ in the proportional part. The feedforward gain is calculated as the ratio between the static gains in P_1 and P_2 and is $K_f = -K_{p2}/K_{p1} = -4$.

Figure 6.26 shows the control in the event of a setpoint change in the flow. When the setpoint change is made, it is the fast control loop with the controller output u_1 that reacts faster and stronger. It ensures that the flow quickly reaches the new setpoint. The slow control loop then makes sure that the signal u_1 is eventually steered back to the desired level $r_u = 50$ %. □

Mid-Range Control Based on Feedforward

Mid-range control using a valve position controller works well in many cases, but there are limits to this method. A problem is related to Example 2 in Figure 6.23. Assume that the valve position controller has a setpoint $r_u = 90\%$ and that you want to increase the flow significantly. This means that the controller output of the flow controller is constrained when it reaches 100%. Since the control error in the valve position controller is only $90 - 100 = -10\%$, it will take a long time for the pump to increase the flow so that it reaches the setpoint.

Another disadvantage of using a valve position controller is that the flow controllers in the two examples cannot be put into manual mode without breaking up the flow control loop. If the flow control loop is in manual, the valve position controller cannot steer the flow to the correct level even though it controls the flow. This is because the VPC does not have access to the flow measurement. A mid-range strategy that does not have these two disadvantages is shown in Figure 6.27.

As with the valve position control, the process section P_1 and the controller C_1 form a fast and accurate control loop. The controller C_2 is no longer a valve position controller, but just like C_1 is also a flow controller with the same setpoint and process variable as C_1. Controller C_2 is a P-controller without an integral part to avoid stick-slip motion caused by friction. The mid-range adjustment of u_1 is done via feedforward to C_2. For this reason, the structure in Figure 6.27 is called *Feedforward Mid-Range Control* (FFMRC).

The feedforward signal u_3 is calculated as follows. To avoid stick-slip motion due to friction, let the controller output u_1 first pass through a dead zone. The user may enter two parameters u_{low} and u_{high} for the dead zone, which define an acceptable stationary value range of u_1, $u_{low} \le u_1 \le u_{high}$. This means that you do not control

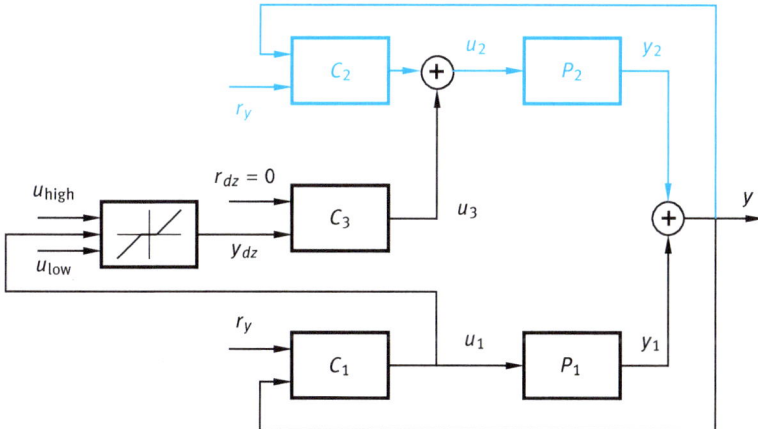

Fig. 6.27: Mid-range control based on feedforward (FFMRC).

u_1 to a certain reference value r_u, but to a region that is acceptable for u_1 when the control loop is in steady-state. If you really want to control u_1 to a reference value r_u, you can choose $u_{low} = u_{high} = r_u$, but then you run the risk of getting stick-slip motion when the valve in process P_2 wears out. The output from the dead zone, y_{dz}, is

$$y_{dz} = \begin{cases} u_1 - u_{high} & u_1 > u_{high} \\ 0 & u_{low} \le u_1 \le u_{high} \\ u_1 - u_{low} & u_1 < u_{low} \end{cases}$$

This output signal is sent to a third controller, C_3, with a setpoint of zero. The controller output signal u_3 from C_3 is the feedforward signal sent to C_2. The controller C_3 is a PI-controller and just like the P controller C_2, it should be tuned conservatively so that it does not disturb the fast control loop too much. The following example shows that the feedforward strategy solves the problem that the valve position control has when limiting the controller output from the flow controller.

Example 6.11. Mid-range control with VPC and FFMRC
Figure 6.28 shows a simulation of the problem corresponding to Example 2 in Figure 6.22 where the setpoint for the controller output from the flow controller is $r_u = 90\%$. The figure shows control with both valve position controller (VPC) and the method based on feedforward (FFMRC).

The processes are the same as in Example 6.10 as are the controller settings for the VPC method. There is no dead zone entered in the FFMRC method, that is, $u_{low} = u_{high} = r_u$.

The controller C_1 in the FFMRC method is tuned in the same way as for the VPC method. The controller C_2 has the gain $K = 3.2$ and the setpoint weight $b = 1$. The controller C_3 has the parameters $K = 0.3$, $T_i = 20$ s and $b = 0$.

Figure 6.28 shows that the VPC method gives a slow control when u_1 is constrained, while the FFMRC method gives a much faster control. In particular, compare the controller outputs u_2 which correspond to the pump speed and shows a sharp increase for the FFMRC method that results in a quick adjustment to the new setpoint. □

6.7 Split-Range Control

Split-range control means that the controller has two or more actuators for a single controller output. The current size of the controller output determines which of the actuators is to be used. The method can, for example, be used for processes where there are possibilities for both heating and cooling. The operating range of the controller output can then be divided, for example, so that controller outputs between 0 % and 50 % control a coolant, while signals between 50 % and 100 % control a heater. This is illustrated in Figure 6.29. The cooling should then be at its maximum setting for con-

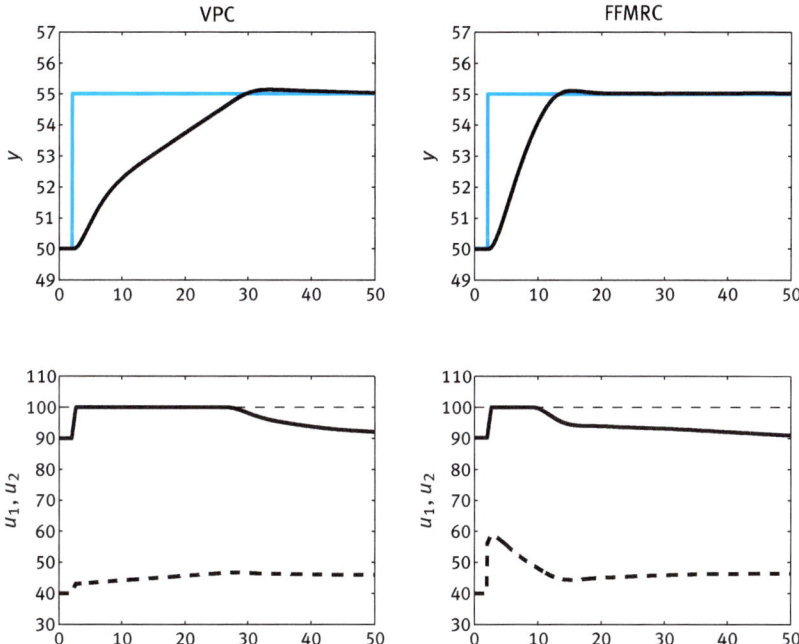

Fig. 6.28: Mid-range control in Example 6.11 with the VPC method (left) and the FFMRC method (right). The upper figures show the flow and its setpoint. The setpoint changes at time $t = 2$ s from 50 % to 55 %. The lower figures show the controller outputs u_1 (solid line) and u_2 (dashed line).

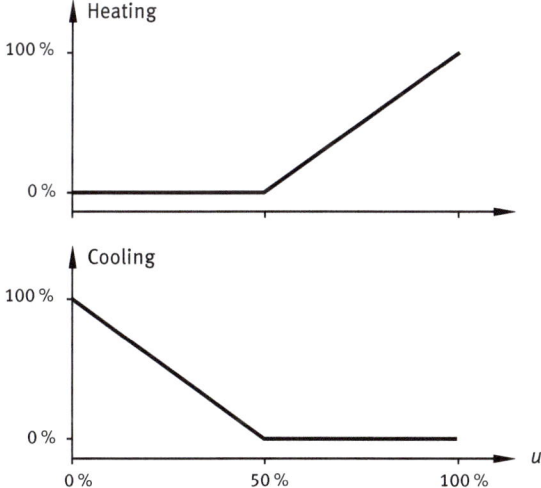

Fig. 6.29: Temperature control with split range.

troller output 0 % and the heating at its maximum setting for controller output 100 %. This may require that signals are reversed. It is possible to have overlapping areas, so that both heating and cooling are active around the turning point of 50 %. Preferably, this should be avoided because it causes unnecessary energy losses, but can be a way to guarantee that no dead zone occurs in the controller output at the turning point.

Example 6.12. Split-range control

Figure 6.30 shows the simulation of a process where you can both heat and cool, for example the temperature in a water pipe. At the start, the heat flow is switched off and a cooling flow of 80 % is needed to keep the water temperature at its setpoint. At

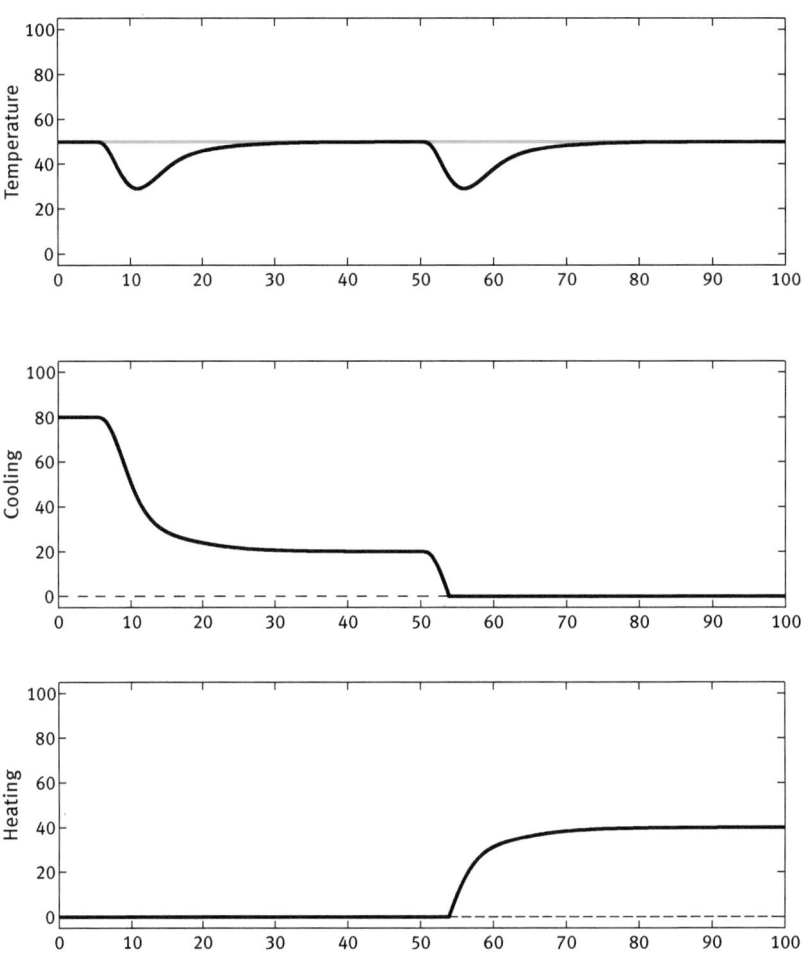

Fig. 6.30: Temperature control with split-range in Example 6.12. The top figure shows the temperature and its setpoint. The middle figure shows the cooling flow and the bottom figure shows the heat flow. The control is disturbed by two load disturbances at the times $t = 5$ s and $t = 50$ s.

time t = 5 s a load disturbance occurs which decreases the temperature. As a result, the cooling flow is reduced to 20 %. At time t = 50 s another load disturbance occurs which decreases the temperature even further. Now it is no longer enough to decrease the cooling flow. Rather, the cooling flow is switched off completely and the heat flow is introduced and increases to 40 % to raise the water temperature to the setpoint.

The controller in the example is a PI-controller tuned with the AMIGO method. It is important to tune the controller to avoid an overshoot in the controller output because an overshoot can result in unnecessary switching between the two flows. ☐

Since split-range control means that the process changes when you change the output from the controller, you usually get a process with varying dynamics. You can make the process gains equal in size by adjusting the turning point from 50 % to another value. However, the best way to compensate for the nonlinearity is to use different controller parameters in the two regions. This can be achieved by using gain schedul-ing as described in Chapter 5.

Another common application of split-range control is concentration control or neutralisation. In these cases, we often have one output that controls the supply of an acidic medium and another output that controls the supply of an alkaline medium. Here, too, overlapping areas should be avoided, as an overlap means that we consume an unnecessary amount of the added media. In both examples described here, it is important that the control loop is well damped. Fluctuations involve direct costs: for temperature control in the form of energy losses and for concentration control in the form of increased consumption of chemicals.

Another area of application for split-range control is the use of several pumps to control a single, common flow. At low flow rates, it may be sufficient for one pump to be running. As the desired flow increases, more pumps may be required to be active. This can be solved effectively with split-range control.

6.8 Dead Time Compensation

Long dead times make it difficult to control dynamic processes. When describing pro-cess models in Chapter 2, we defined the dead time as the time it takes from making a change in the controller output until a reaction is seen in the process variable.

Occurrences

There are essentially two causes of dead time in processes. The first cause is material transports in the process, the second cause is dead time caused by delays in computers and networks.

Transport Delays

Dead times often occur in connection with material transports, in pipes or on conveyor belts. Flow and level control of liquids are normally free of dead times, as a change in the valve position momentarily results in a change in the flow in the entire pipeline. Concentration control of liquids, on the other hand, often includes dead times. This was illustrated in Example 6.7. If the concentration sensor is located at a distance l m from the mixing valve and the flow is f m/s we get a dead time that is

$$L = \frac{l}{f}$$

In other words, a large distance and a low flow give a long dead time.

Flow and level control of solid materials transported on belts also often results in dead times. A flow change is not registered momentarily, but only after the material has been transported to the sensor, for example a scale. If the scale is l m from the position where the material is placed on the belt and if the belt has the speed v m/s, the dead time will be

$$L = \frac{l}{v}$$

A large distance and a low speed thus gives a long dead time.

Sometimes the dead time can be reduced by moving the location of sensors and actuators. When designing your process, you should think about placing sensors and actuators so that you get dead times that are as short as possible.

Computers and Networks

Most PID-controllers today are implemented using computers. This has certain consequences that we will briefly discuss here.

Today's computer-based controllers have mostly only advantages when comparing them to yesterday's analog electrical or pneumatic controllers. Computers make it much easier to add more advanced functions such as automatic tuning, alarm handling, filtering, logic and sequence control. It is also possible to set parameters more accurately. The function of the controller does not change in the same way as before due to aging components.

However, there is an important disadvantage. Today's controllers do not process the signals analogously, but the signals are sampled, that is, they are measured and recorded at regular intervals. Between measurements, the controller does not receive any new information about the sensed value. This is not a problem as long as the time interval between the samples is much shorter than the process time constants. However, with fast control loops, for example pressure and flow control loops, one should be careful.

Sampling means that we introduce an extra dead time in the control loop. Figure 6.31 shows a continuous signal together with its sampled form. The figure clearly shows that the sampled version compared to the actual signal is delayed. The delay

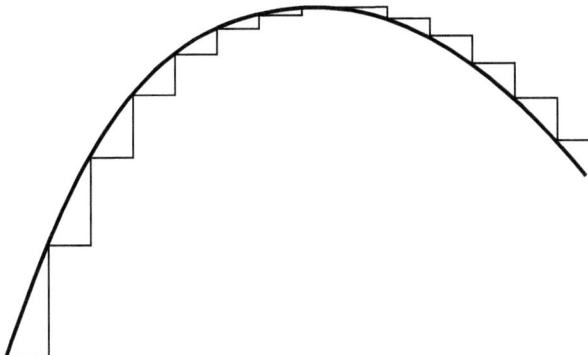

Fig. 6.31: A continuous signal with its sampled realisation.

is on average about half the sampling period. In most controllers, the new controller output is not sent out immediately after each new measurement. Instead, you wait until the next measurement. This way of handling the calculations results in a delay of another sampling period in the control loop. The total delay that arises due to the sampling is therefore one and a half sampling periods. This delay can cause problems if the process is fast.

In most single-loop controllers, the sampling period is fixed; in process controllers the sampling period is between 0.1 s and 0.3 s, in temperature controllers often longer. In industrial control systems, on the other hand, there is often an option that allows you to choose the sampling period. A rule of thumb when choosing sampling periods in controllers is that the sampling period should be shorter than about one tenth of the dominant time constant of the closed loop. If you cannot choose the sampling period, you must accept a control loop that is so slow that it meets the requirement.

Some control systems are designed so that each individual function block can give rise to a delay corresponding to one sampling interval. Here it is particularly important to check that the total delay does not become too long. In recent years, processes have started to be controlled over networks. This means that additional dead times are added. In many networks it is not known how long the dead time is, but it varies according to the load on the network. This can lead to major problems. Control over such networks should be avoided, at least in the case of critical processes.

Some control systems use dead bands on process variables. This has the effect that you do not need to update measurement values if they have not changed very much. The reason for doing this is that you want to reduce the load on your computer. From a control point of view, however, this is a very bad way of reducing the load on the computer, as this introduces dead times in the control loops. If possible, you should therefore avoid this function and if necessary reduce the load on the computer in other ways.

PID-Control of Dead Time Processes

When describing the PID-controller in Chapter 3, we saw that the D-part is used to predict future control errors. It does this by considering how the process variable changes, that is, by examining its derivatives. This is an effective method when we have higher order processes, but the method is not good when we have long dead times in the process.

Assume, for example, that we have made a setpoint change, which results in the PID-controller starting to control the process towards the new setpoint. If the dead time is long, a good controller should perform a large part of its control operation during the dead time. During this time, the process variable has not had time to react at all to the new control actions. There is therefore no information in the process variable that the D-part can use for its prediction. The D-part will therefore only be a nuisance and disrupt the control when the dead time is long.

If you are going to control a process with a long dead time with a PID-controller, you should therefore remove the D-part and only use a PI-controller. Chapter 4 described different ways to tune PI-controllers. It was mentioned that Ziegler-Nichols tuning rules do not work very well when we have long dead times, but that a better tuning rule is

$$K = \frac{K_c}{4}$$
$$T_i = \frac{T_c}{4}$$

where K_c is the critical gain and T_c is the critical period time. This tuning means that the controller works less with the proportional part and more with the integral part than with the usual tuning rules. The method is consistent with the AMIGO method when the process has a long dead time. If you want to use step response methods for processes with long dead time to tune the controller, the AMIGO method or the One-third rule are recommended.

Otto-Smith Controller

You can of course control long dead time processes with a standard PI-controller, but since we have removed the derivative part this results in a controller that does not predict the future. This is a major disadvantage, because a prediction of future control errors is very helpful — particularly when we have long dead times. Therefore, special controllers have been designed for processes with long dead times that can predict future control errors, but without using the derivative of the process variable.

The principle of dead-time controllers is that instead of considering changes in the process variable during the prediction, a process model is generated and the controller output is the input signal to the model. This simulates the process variable and thus

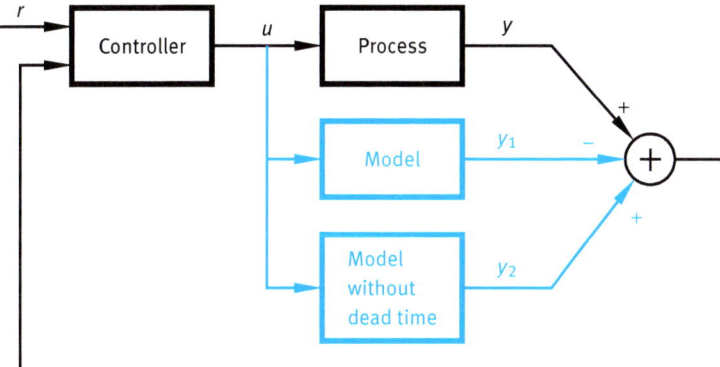

Fig. 6.32: Principle of an Otto-Smith controller.

the control error in the near future. The most common dead-time controller is the Otto-Smith controller. The principle behind it is shown in Figure 6.32.

In the Otto-Smith controller there is an internal controller which is usually a PI-controller. The output signal from this controller acts on the process as usual, but it is also an input signal to a model of the process as well as to a model of the process with the dead time removed.

For the sake of simplicity, let us assume that the model is an exact description of the process and that there are no disturbances. The two signals y and y_1 will then be identical, and therefore completely cancel each other out. What remains as a signal to the internal controller is then the signal y_2, that is, the process variable we would have received if we had not had dead time in our process. In this way, the controller will work towards a simulated process that is identical to the actual process except that the dead time is removed. The control will also be as good as it would have been if we had no dead time, apart from the fact that the process variable y is of course still delayed.

In practice, of course, the model is not an exact description of the process. This means that it is not exactly y_2 that goes back to the controller. This in turn means that you normally have to tune the controller more conservatively than you would have done if there was no dead time.

An Otto-Smith controller is usually tuned as follows:
1. Perform a step response experiment on the uncontrolled process. Determine the process dead time, time constant and process gain from the step response and enter these values in the controller.
2. Tune the PID parameters of the controller. Ideally, you can tune them as if there was no dead time. Disturbances and uncertainty in the model, however, mean that controller tuning must be more careful and conservative.

The following example shows how the Otto-Smith controller works.

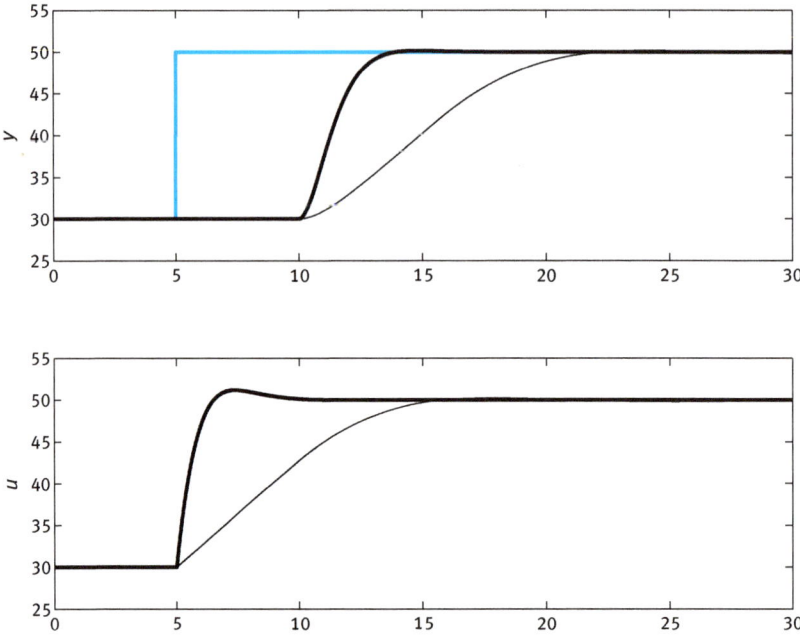

Fig. 6.33: Comparison between the Otto-Smith controller (thick lines) and a PI-controller (thin lines). The figure shows the control in the event of a setpoint change.

Example 6.13.   Otto-Smith controller

Here, a simulated example shows how the dead time compensation ideally works. Figure 6.33 shows the control of a first order process with a time constant of one second with an additional dead time of five seconds. The figure shows both the result of control with a PI-controller and with an Otto-Smith controller.

The PI-controller is tuned so that it provides as fast control as possible without overshoot. The controller parameters are $K = 0.3$ and $T_i = 2.35$ s. When the setpoint change is made, the controller begins to carefully integrate the control error. The integration takes place so slowly that no overshoot occurs. The dead time controller is also tuned so that it does not give an overshoot. The controller parameters are $K = 1.0$ and $T_i = 0.7$ s, which means that the gain is significantly larger and the integral time significantly shorter compared to the previous PI-controller. The Otto-Smith controller performs its entire control operation before the process variable has even begun to react.

In the event of load disturbances, none of the controllers has any opportunity to act until the disturbance is visible in the process variable. Here, only feedforward of the disturbance can significantly improve the result. □

PPI-Controller

Although the Otto-Smith controller provides significantly better control of processes with long dead times, it is still the PI-controller that is by far the most common controller for these processes. The main reason for this is, of course, that people are used to handling the PI-controller and that in many cases it is considered to provide sufficiently good control. Another reason is the complexity of the Otto-Smith controller. In the previous example, the Otto-Smith controller contained a process model comprising three parameters: dead time, time constant and process gain. Together with the PI-controller's two parameters, this means that no less than five parameters must be tuned.

The PPI-controller is a dead-time compensating controller which is identical in structure to the Otto-Smith controller, but which is similar to a PID-controller in terms of handling. The name PPI stands for Predictive PI-controller. What distinguishes the PPI-controller from other dead time compensating controllers is that the user does not have to specify a process model. The controller itself determines the process model from the controller parameters specified by the user.

The equation for the PID-controller is

$$u(t) = K\left(e(t) + \frac{1}{T_i}\int e(t)dt + T_d\frac{de(t)}{dt}\right)$$

where the third term, the derivative term, is used for prediction. The three parameters in the PID-controller are the gain K, the integral time T_i and the derivative time T_d.

The equation for the PPI-controller is

$$u(t) = K\left(e(t) + \frac{1}{T_i}\int e(t)dt\right) + \frac{1}{T_i}\int (u(t - L) - u(t))dt$$

The three parameters of the PPI-controller are the gain K, the integral time T_i and the dead time L. The first two terms are identical to those of the PID-controller, but the third term in the PPI-controller has been replaced by a term that consists of the integral of the difference between the controller output we had L seconds ago and the current controller output. In other words, the third predictive term is based on the controller output instead of on the process variable.

Tuning of PPI-Controller

The easiest way to tune the PPI-controller's three parameters is to first perform a step response experiment on the process and from this determine the process gain K_p, dead time L and dominant time constant T. The three parameters of the PPI-controller can then be tuned according to the following simple rules of thumb:

$$K = \frac{1}{K_p}$$

$$T_i = T$$

$$L = L$$

The PPI-controller can then, if necessary, be adjusted manually according to the rules of thumb for the PI-controller described in Section 4.3. For example, an increase in the gain or a decrease in the integral time provides in most cases a faster control, but at a price of decreased robustness.

6.9 Ratio Control

In ratio control we want to control the ratio between two process variables. In the case of combustion, for example, the ratio between the fuel supply and the air supply should be controlled so that the combustion is as efficient as possible. In the case of chemical mixing processes, we want to mix two or more flows in given proportions. For the time being, we will stick to the most common case: controlling the ratio between two flows.

Ratio control means that we have a control problem with two process variables, the two flows, two controller outputs that control the two valves or pumps, and two setpoints that correspond to the two flows. These two setpoints are in turn calculated from two higher-level setpoints, namely the desired flow, either the total flow or the flow in one of the lines, and the desired ratio between the two flows. Ratio control can be performed in several different ways. We start with the most common methods, namely those that are based on using a ratio station.

Ratio Station

Figure 6.34 shows a very common way to solve the ratio control problem, namely by using a parallel ratio station. There are two simple control loops. The upper loop consists of process P_1 and controller C_1 and the lower loop of process P_2 and controller C_2. The process variables y_1 and y_2 are the two flows that we want to be kept at a ratio denoted a, that is, we want to achieve that

$$\frac{y_2}{y_1} = a$$

You not only want to control the ratio between the two flows but also the flow itself. We assume here that you do this by giving a flow setpoint r to the upper control loop,

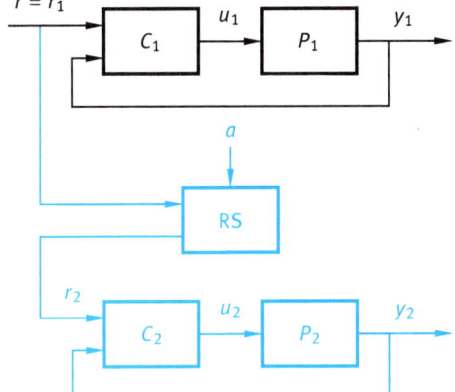

Fig. 6.34: Ratio control with a parallel ratio station.

that is, by setting $r_1 = r$. If you instead want r to denote the total flow $y_1 + y_2$, you can instead choose $r_1 = r/(1 + a)$.

Figure 6.34 illustrates how ratio control is performed using a ratio station, or RS for short. A ratio station is simply a multiplier. This means that the setpoint r_2 in the lower loop is calculated as

$$r_2 = ar_1$$

You take the setpoint of the upper loop, multiply it by the desired ratio a and pass this signal to the setpoint of the lower loop. If both controllers contain integral parts, this will mean that $y_1 = r$ and $y_2/y_1 = a$ when the control is in steady-state.

Using a parallel ratio station thus works when the control is in steady-state and there are no disturbances. If you make a setpoint change, the ratio will also be kept if the two control loops have approximately the same dynamics. You can ensure that they have this by tuning the controllers appropriately or by low-pass filtering the set-point of the faster of the two loops. However, there is no attempt to keep the ratio in case of load disturbances or when either of the two controller outputs is constrained.

Figure 6.35 shows how to solve the ratio control with a serial ratio station. The difference is that you use the actual flow y_1 as an input signal to the ratio station instead of the setpoint r_1, that is

$$r_2 = ay_1$$

This means that the lower loop follows the upper loop. The upper loop is therefore usually called the primary loop or master loop, while the lower loop is called the secondary loop or slave loop. The advantage of the serial ratio station compared to the parallel one is that you now act on disturbances in the upper loop. If there is a load disturbance in the upper loop or if its controller output is constrained so that the flow is no longer at the setpoint, the lower loop will act to try to maintain the desired ratio between the flows. This also applies if you put the controller C_1 in manual control.

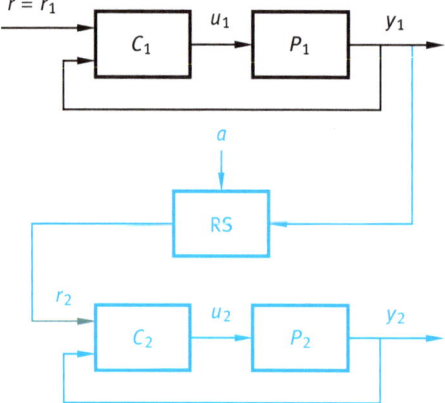

Fig. 6.35: Ratio control with a serial ratio station.

However, the problem of disturbance or controller output constraints still remains in the lower loop. The serial ratio station does not take care of these. A disadvantage compared to the parallel ratio station is that you do not manage to keep the ratio as quickly when encountering setpoint changes because the flow y_2 will be delayed compared to setpoint $r_2 = ay_1$. The length of the delay is determined by the dynamics of the secondary loop. This is shown in the example below. Another advantage of the parallel ratio station is that measurement disturbances in the upper loop are not passed on to the lower loop through its setpoint.

Example 6.14. Ratio control with serial ratio station

Figure 6.36 shows the simulation of the ratio control in Figure 6.35. Both processes are simulated with two first order processes in series. The time constants in process P_1 are 5 s and in process P_2 the time constants are 1 s. Both controllers are PI-controllers. The controller C_1 has the parameters $K = 1.5$ and $T_i = 6.7$ s and the controller C_2 has the parameters $K = 1.5$ and $T_i = 1.3$ s. The desired ratio between the two flows is selected to $a = 1$.

The figure illustrates the use of ratio control for two setpoint changes. In steady-state, the flow in the primary loop is equal to its setpoint and the ratio between the two flows is one, that is, it corresponds to the desired ratio $a = 1$. For setpoint changes, on the other hand, the flows deviate from the desired ratio during the transient phases because the flow of the secondary loop is delayed relative to the flow of the primary loop. □

There are cases where it is important that you never get a shortage of one of the two flows. Assume that Example 6.14 describes a combustion process where the primary loop is the fuel control and the secondary loop the air control. In the first setpoint change when the setpoint increases, you then get a deficit of air during the transient phase, while you get an excess of air when the setpoint decreases in the second set-

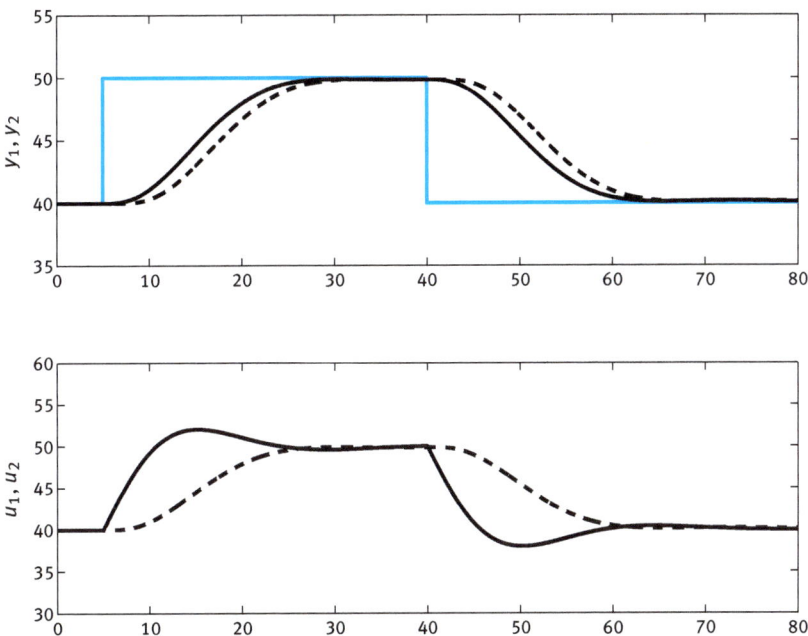

Fig. 6.36: Ratio control with serial ratio station in Example 6.14 with two step changes in the set-point. The solid lines show the signals from the primary loop and the dashed lines show the signals from the secondary loop.

point change. If you want to prevent a deficit of air that would result in fuel not burning completely, you should let the external setpoint determine the air control when the setpoint *increases* and determine the fuel control when the setpoint *decreases*. In other words, the roles of the primary loop and the secondary loop shift between the two loops. This can be solved using MIN and MAX selectors as shown in Figure 6.37.

Example 6.15.   Ratio control with ratio station and selectors

Figure 6.38 shows a simulation of ratio control with selectors as described in Figure 6.37 applied to the same processes and controllers as in Example 6.14.

The figure shows that the steady-state flows correspond to the desired ones but that the desired ratio cannot be kept during setpoint changes. One difference from the simple ratio station is that we have now managed to avoid a deficiency in air. The air flow y_2 is never lower than the fuel flow y_1, which means that the ratio between the two flows, y_2/y_1, is never less than the desired one, in this case $a = 1$.

In Figure 6.38 it can be seen that the deviation between the two flows becomes particularly large as the setpoint increases. The reason is that the air loop is significantly faster than the fuel loop and in this case the air loop is the primary loop. If this is a problem, you can make the air loop slower or let the setpoint of this loop pass through a low-pass filter or a ramp function. □

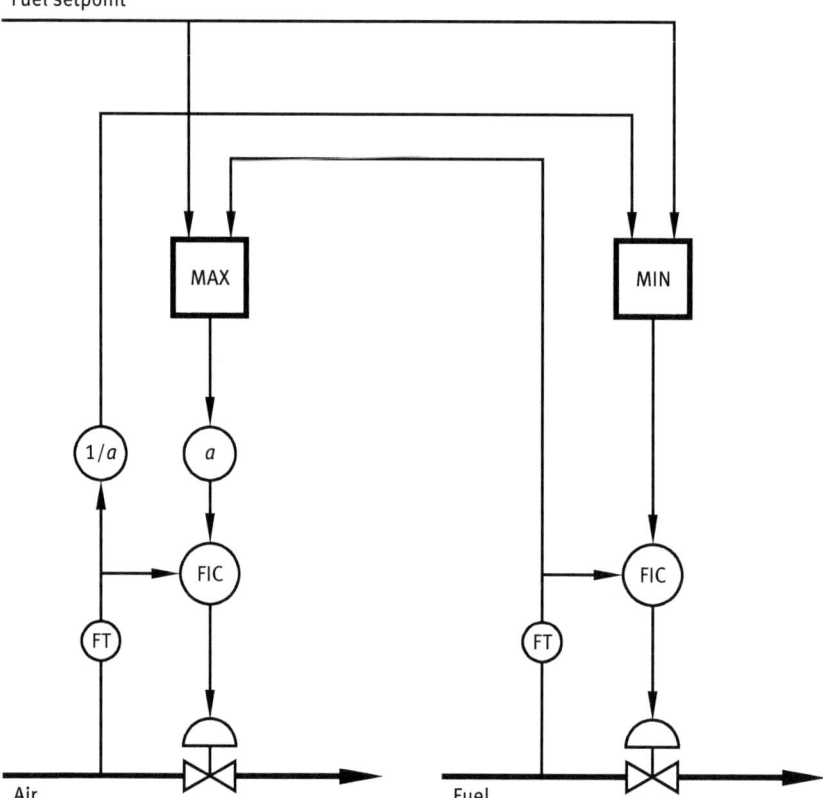

Fig. 6.37: Ratio control of the fuel/air ratio with the help of selectors.

The solution according to Figure 6.37 is also advantageous from a safety point of view. If the air supply should for some reason decrease or cease completely, there is a risk of explosion in the system. Through the solution in Figure 6.37, the fuel controller will automatically restrict the fuel supply if the air supply should cease.

Ratio control is not only used in mixing processes to control the ratio between flows. In the following example, you want to keep the ratio between a flow and a temperature constant.

Example 6.16.   Control of flow through a pasteuriser
In a pasteuriser, a liquid must be heated to a specific temperature and be kept at this temperature for a certain minimum time. Figure 6.39 shows the control of the flow through a pasteuriser.

A liquid is heated in the pasteuriser and then transported to a buffer tank. The liquid is emptied from the buffer tank and the flow of this outlet varies. The control problem is to keep the level in the buffer tank within certain limits despite the varying outlet flow.

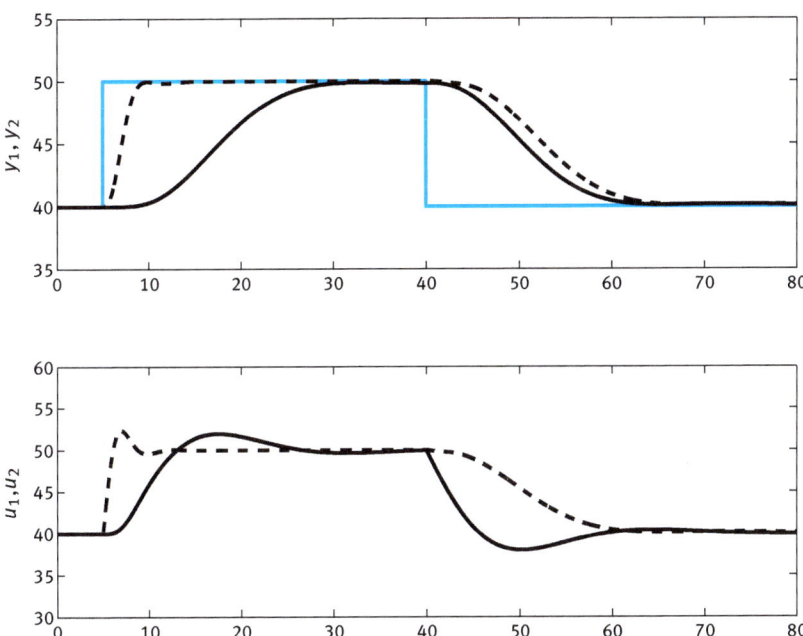

Fig. 6.38: Ratio control with ratio station and selector in Example 6.14 for two step changes in the setpoint. The solid lines show the signals from the fuel loop and the dashed lines show the signals from the air loop.

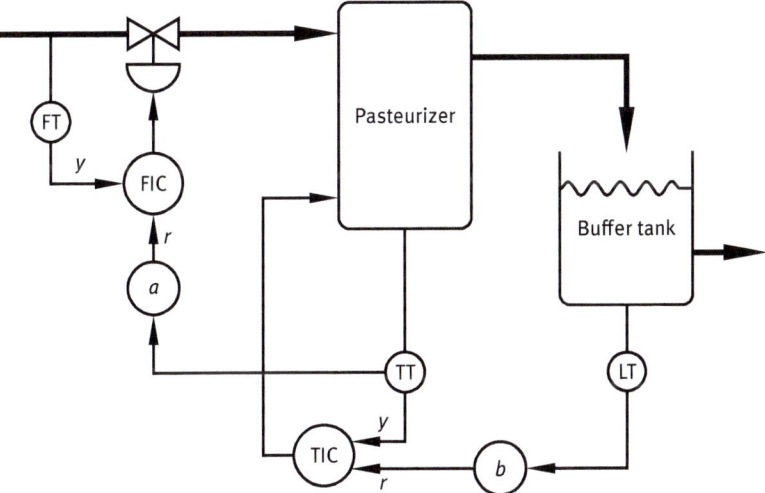

Fig. 6.39: Control of the flow through a pasteuriser with ratio control.

It is important that the liquid in the pasteuriser is kept at the high temperature for a sufficiently long time. When you want to increase the outflow, you must therefore first increase the heat in the pasteuriser. If you would momentarily increase the flow through the pasteuriser, the outflow would cause a temperature drop.

Therefore, the adjustment is suitably performed as shown in Figure 6.39. Assume that the outflow from the buffer tank increases. Then the level in the buffer tank drops. Instead of momentarily increasing the flow through the pasteuriser, you increase the setpoint of the temperature controller. As the relatively slow temperature control causes the temperature in the pasteuriser to rise, the flow to the pasteuriser is increased by the ratio controller. The constant a indicates the relationship between the temperature and the flow. As the temperature rises, so does the flow. This in turn leads to the level in the buffer tank being steered back to the correct level.   □

Tracking Ratio Station

The most common way to solve ratio control problems is to use the parallel or serial ratio station. However, these methods have their limitations. The parallel ratio station does not take into account load disturbances and situations in which any of the controller outputs are constrained. The serial ratio station is unable to maintain the ratio in the event of setpoint changes. It handles load disturbances and constraints of the controller output of the primary loop, but not disturbances and limitations in the secondary loop. The tracking ratio station (TRS) is a strategy that is able to follow the ratio in case of setpoint changes, load disturbances and controller output limitations in both loops.

In Example 6.15, selectors were used to switch the roles of primary loop and secondary loop between the different control loops to ensure that we never experience a shortage of air during combustion. The switching was determined by the direction of the setpoint change. In the following ratio station, selectors are also used to change the roles as primary and secondary loop, but here the size of the control errors in the two loops is used to determine the switching. The principle is that the control loop that has the largest control error becomes the primary loop, while the loop that has the smallest control error becomes the secondary loop and follows the process variable of the primary loop. With this method, setpoint changes, load disturbances in both loops and also controller output limitations in both loops will be taken into account. It is also possible to put one of the controllers in manual mode and still follow the ratio a.

The functionality of the tracking ratio station is described in Figure 6.40. The ratio station has four input signals, the two flows y_1 and y_2, the flow setpoint r and the desired ratio a. Based on these four inputs, the ratio station determines the setpoints r_1 and r_2 for the two loops. First, the control errors e_1 and e_2 are calculated for the two loops. The absolute values of the two errors are then compared, after which a selector

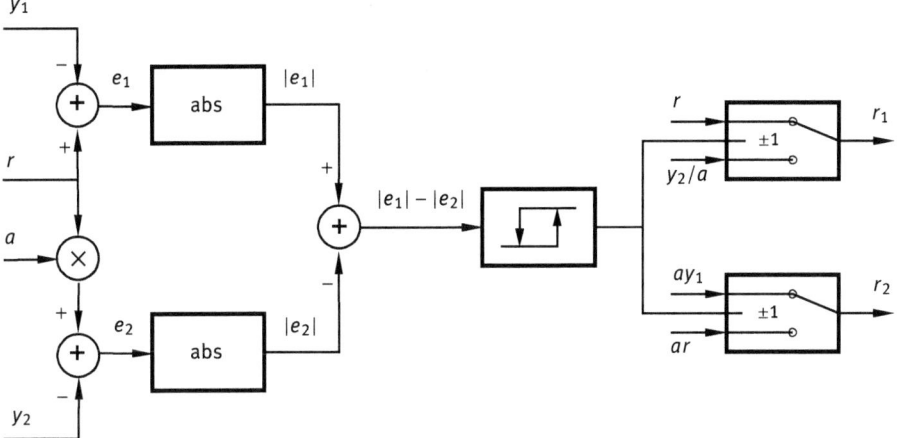

Fig. 6.40: Tracking ratio station (TRS).

function determines which loop should be the primary loop and follow the external setpoint and which loop should become the secondary loop and follow the process variable of the other loop. A hysteresis is inserted to prevent too frequent switching between the primary loop and the secondary loop.

The diagram in Figure 6.40 is relatively complicated to build up with function blocks, but the tracking ratio station has begun to be implemented as a separate function block in modern control systems and the function thus becomes easier to use. The following example illustrates the operation of the tracking ratio station.

Example 6.17. Ratio control with tracking ratio station
Figure 6.41 shows a simulation of ratio control with the tracking ratio station with the same processes and controllers as in Examples 6.14 and 6.15.

The figure shows the response to a setpoint change followed by a load disturbance. The tracking ratio station manages to keep the ratio between the two flows constant during both changes.

The price you have to pay for this careful ratio control is switching of the controller outputs, which can wear out the control equipment quickly. If you can not accept this, you can introduce a larger hysteresis in the ratio station or low-pass filter the setpoints. However, this means that you can not keep the desired ratio as well. This is a compromise that can only be resolved on a case-by-case basis. □

In this section, we have considered the desired ratio a as a constant, but you do not always want to keep it constant. During combustion, for example, the ratio is often allowed to vary depending on measurements of the oxygen concentration in the exhaust gases in an attempt to optimise the combustion. In ratio control, we have two setpoints we want to follow: the flow and the ratio between the contributing sub-flows. In this

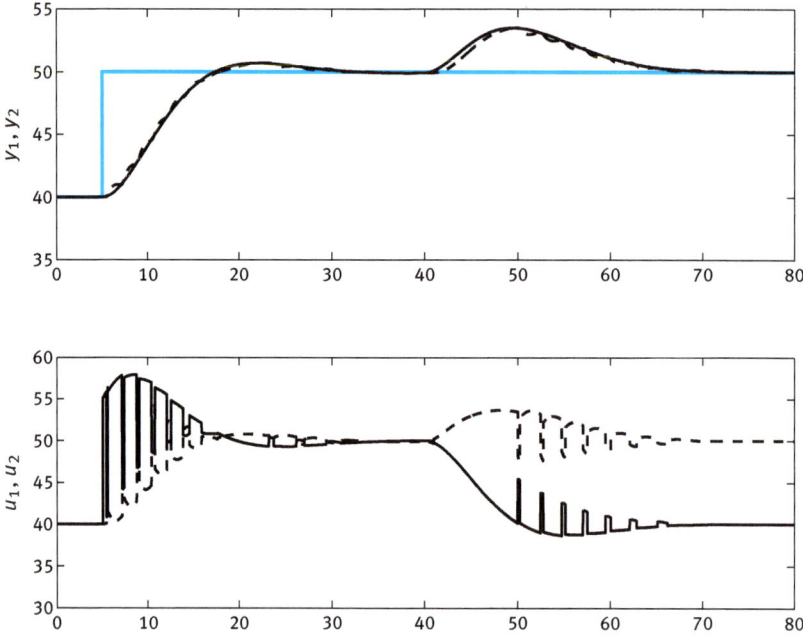

Fig. 6.41: Ratio control with the tracking ratio station in Example 6.17. At $t = 5$ the setpoint changes and at $t = 40$ there is a load disturbance on the process P_1. The solid lines show the signals from the upper loop and the dashed lines show the signals from the lower loop.

section we have assumed that the ratio is the most important variable and we let the flows vary to try to keep this ratio. In many cases, you have the reverse problem, that is, the primary objective is to keep one of the flows at its setpoint while the ratio is less important. In these applications, it is recommended to use the serial ratio station and let the important flow belong to the primary loop.

6.10 Decoupling

Controlling an industrial production process comprises many individual control loops. These control loops are connected in various ways, including using the methods described in this chapter. In most cases, these connections are desirable and deliberately made, but there are cases where control loops affect each other in an undesirable way. If this happens it may be necessary to use filters to decouple control loops.

Consider a wash basin with a hot water tap and a cold water tap. We want to control both the temperature and the flow with these two taps. This is an example where two controller outputs affect both process variables. If, for example, we want to increase the water temperature, we can do so by opening the hot water tap further.

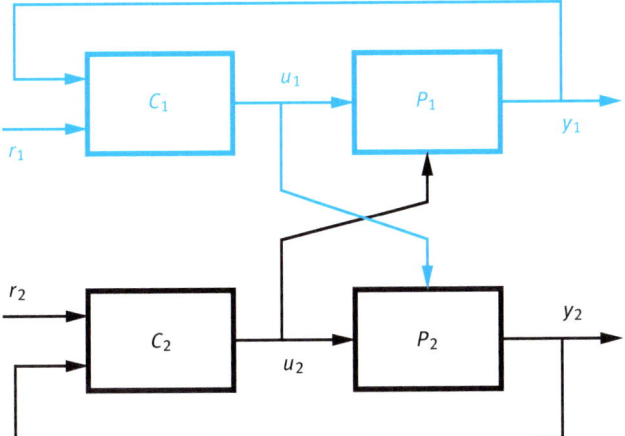

Fig. 6.42: Two coupled control loops. The controller outputs from the two controllers act as load disturbances to the other loops.

However, this will mean that the flow will also increase. If we want to raise the temperature while maintaining the flow, we must close the cold water tap just as much as we open the hot water tap. The problem of coupled control loops is illustrated in Figure 6.42.

We have two controllers that each control a process section, but the controller outputs from the controllers also affect the other control loop. We can see the two controller outputs as load disturbances in the other loops.

In the process industry, the problem of coupled loops is usually solved as follows. First, you prioritise and decide which of the two loops is more important. The controller of this loop is tuned as usual. Then you tune the second loop controller, but do so conservatively resulting in a slowly varying controller output that affects the first loop only slightly. This way of handling the coupling between the control loops often works, but it is inefficient because the second loop is tuned unnecessarily slow.

A better way of solving the problem is to use decoupling. The method is illustrated in Figure 6.43. Each controller output is coupled to the opposite controller via a feedforward filter. The filters F_1 and F_2 can be gains, but using lead/lag filters gives a more efficient decoupling. Lead/lag filters were described in Section 6.2. and are sometimes combined with delays. The task of the decoupling is to interrupt the connections between the two control loops so that they do not interfere with each other. The controllers can then be tuned and used independently of each other. The method is illustrated in the following example.

Example 6.18. Pressure and flow control of pulp

Figure 6.44 shows the principle of controlling a mass transport from a storage tower to a tank. The control objective is to control the level in the tank. Therefore, level con-

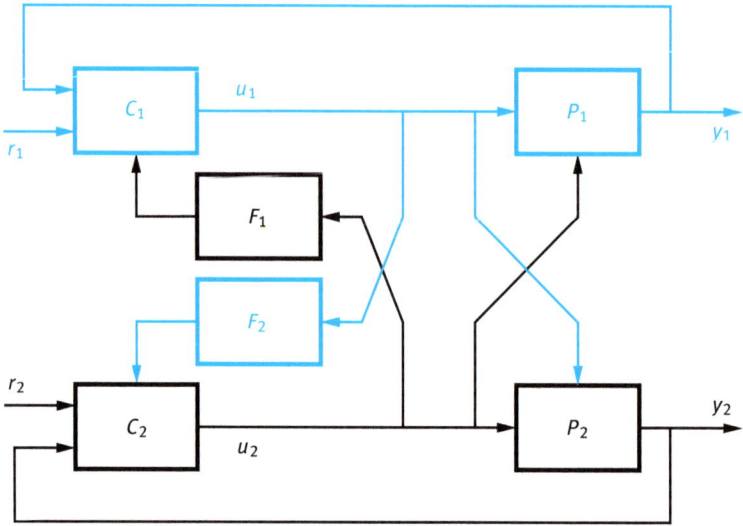

Fig. 6.43: Two coupled control loops where the coupling is reduced by means of feedforward filters F_1 and F_2.

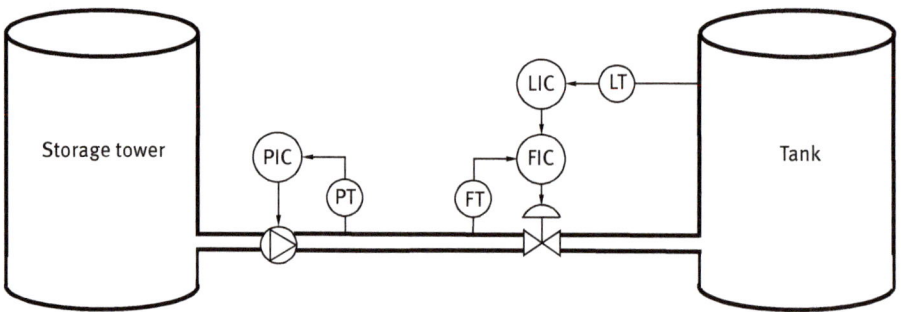

Fig. 6.44: Control of mass transport from a storage tower to a tank.

troller LIC controls flow controller FIC via a cascade loop. Thus, the flow into the tank gives the desired level. Note that the flow controller controls a valve in the pulp line. A pump in the pulp line ensures that the pressure is maintained at a specified level. Therefore, there is also a pressure controller, PIC, which controls the pump.

The flow and pressure control loops are coupled and interfere with each other. For example, suppose you want to increase the flow in the line. Then the flow controller will open the valve. Consequentially, the pressure drops. As a result, the pressure controller will increase the pump speed to increase the pressure. This in turn increases the flow, which means that the flow controller closes the valve. It is easy to see that you can get oscillations in the system if you do not take care of the coupling between the

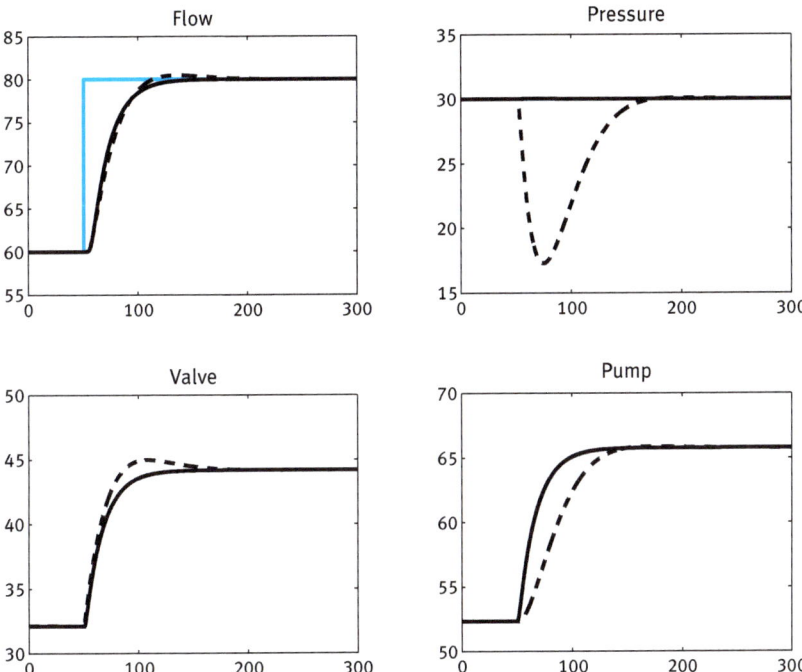

Fig. 6.45: Flow and pressure control in the pulp line in Example 6.18. The solid lines show control with decoupling and the dashed control without decoupling.

control loops. Figure 6.45 shows a simulation of the control both with and without decoupling.

The simulation shows the impact of the two controllers in the event of a setpoint change in the flow. The setpoint change causes the flow controller to open the valve. As a result, the flow increases, but the pressure drops, too. Without decoupling the pressure shows a significant disturbance. The decoupling is realised with feedforward connections according to Figure 6.43. The two feedforward filters F_1 and F_2 are lead/lag filters. With decoupling, the pressure is effectively undisturbed by the change in the flow rate.

The figure shows that to achieve this improvement, no drastic changes to the controller outputs are required. The main difference is that the pump speed increases faster when decoupling and thus prevents the pressure from dropping. □

The example shows that decoupling can be very effective. There are many cases of connected control loops in industry, so decoupling should be used definitely more than done today. The separate coupling can easily be implemented with feedforward functions that already exist in most modern control systems. Of course, effort is required to find suitable feedforward filters, but there are systematic methods for doing this.

If you are content with using only filter gains, you can usually choose them relatively easily from steady-state relationships or by trial and error.

6.11 Buffer Control

A production process normally consists of many different processing stages. Within each processing stage, it is important to achieve effective control, that is, process variables must follow their setpoints. So far, we have studied the control problem of following setpoints but sometimes there is a slightly different control objective. An example of this is buffer control.

The different processing stages are often separated using buffer tanks. This is done primarily so that production breaks and production changes can be handled efficiently. A disturbance in a processing stage then does not necessarily have an effect in adjacent processing stages. Another case where you need buffer tanks is when you have batch production in certain processing stages and connect these to an otherwise continuous production. This is the case, for example, when you have batch reactors in your process. Then you need buffer tanks both before and after the reactors. Buffer tanks are also called balance tanks.

Figure 6.46 shows the principle of controlling a buffer tank when you want to protect the process against operational disturbances that happen upstream, that is, in the process section *before* the buffer tank. With the solution in Figure 6.46, if the upstream flow varies, or even ceases periodically, one can still continue production and maintain a steady flow downstream, that is, in the process section after the buffer tank. In level controlled buffer tanks, the primary goal is not that the process variable follows the setpoint as best as possible. The primary task of the buffer tank is to act as a buffer

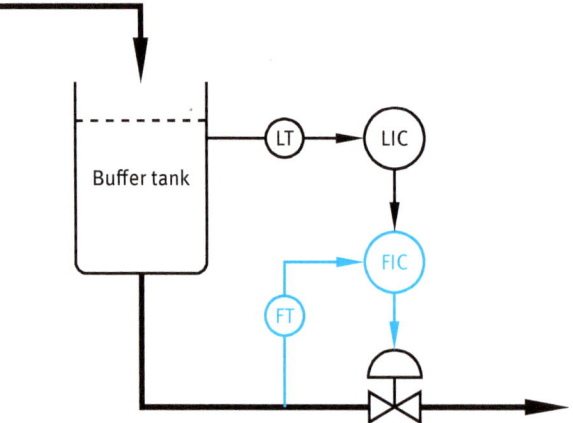

Fig. 6.46: Level control of a buffer tank protecting against upstream disturbances.

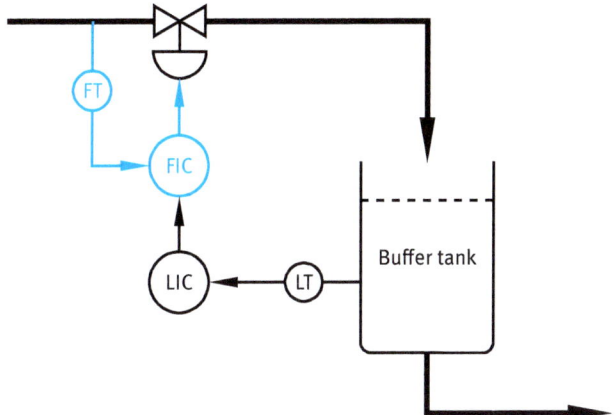

Fig. 6.47: Level control of a buffer tank that protects against downstream disturbances.

between incoming flow and outgoing flow. In control terms, this means that you want to keep the controller output as fixed as possible. The difference between the process variable and setpoint is unimportant as long as the tank does not become empty or overflows. This means that level controller LIC in Figure 6.46 must control cautiously and allow large variations in the level.

Figure 6.47 illustrates the principle for controlling a buffer tank protecting the process against operational disturbances after the buffer tank, that is, downstream. The difference compared to the solution in Figure 6.46 is that here you control the incoming flow to the tank instead of the outgoing flow.

Sometimes the control problem is solved by introducing a large dead zone in the level controller. As long as the control error is within the dead zone, the controller output does not change. This solution means that the controller output will be constant while we are in the dead zone. However, since we have no control during these time periods, the level will rise or fall until the control error becomes larger than the dead zone. The result is an uneven control, with periods of constant controller output interspersed with periods when the controller tries to control the level into the dead zone again.

A better method is to use gain scheduling as described in Chapter 5. Gain scheduling was introduced as a method for handling nonlinear processes. In buffer control, gain scheduling can be used to achieve different types of control in different operating regions. In this case, the process variable, that is, the level in the tank, should be used as the reference signal for the gain scheduling table.

As the level approaches the limits of the tank, you should have a control with relatively large gain so that you quickly leave the "dangerous" area. Within these limits, for example within 20 % to 80 % of the operating region, one should, on the other hand, have a controller with low gain so that the controller output changes as little as possible.

Since we are not interested in achieving the complete elimination of the control error, there is no reason to use the integral part of the PID-controller. In any event, due to the integrating nature of the level process, a P-controller provides significantly more stable control than a PI-controller.

Another good idea for effective buffer control is to vary the setpoint instead of fixing it to 50 %. This is particularly useful for intermittent production. When the production rate is high, the most likely change is that production decreases. For the buffer tank in Figure 6.46, which protects against disturbances upstream, a large setpoint should therefore be selected in order to have a greater capacity to supply the subsequent process step when the supply to the buffer tank decreases. In the same way, at low production, a small setpoint should be chosen, so that the tank can capture the liquid in the event of an increase in the inflow. For the buffer tank in Figure 6.47, which protects against disturbances downstream, the opposite should be done, that is, choose a small setpoint at high production to protect against a reduced outlet when production decreases, and a large setpoint at low production to protect against an increased output as production increases. By changing the setpoint depending on the production level you can increase the buffer capacity in the tank.

Exercises

2. Process Types and Step Response Analysis

2.1 The book describes six different types of processes; first order, higher order, integrating, oscillating, dead time processes and processes with reverse response. In reality, most processes are a combination of several of these process types. Figure E-2.1 shows step responses for six processes that are a combination of two process types. Try to determine what these process types are. The time of the step change in the controller output is marked by dashed lines.

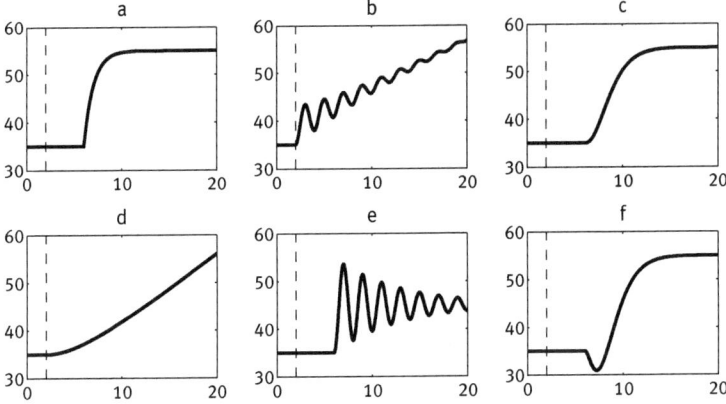

Fig. E-2.1: Step response of six different processes in Problem 2.1.

2.2 You want to determine the process gain using a step response experiment. The controller output was initially 20 % and was then increased to 30 %. As a result, the process variable, which was initially 40 %, increased to 60 %. Determine the process gain K_p.

2.3 A temperature controller should be tuned and it is therefore necessary to derive a model of the process. A step response experiment is performed. At the beginning the controller output was at 30 % and the temperature was 70 °C. The controller output was changed to 40 %. The temperature increased to 85 °C as a result. The range of the process variable was set between 50 °C and 200 °C. Calculate the process gain K_p.

2.4 Figure E-2.2 shows a simulated step response experiment to determine a process model. What process type is shown, that is, which of the six different process types

https://doi.org/10.1515/9783111104959-007

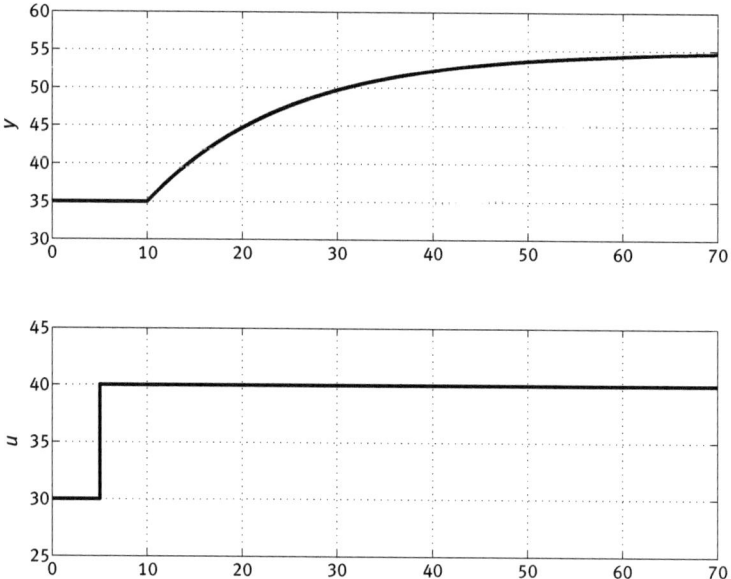

Fig. E-2.2: Step response experiment in Problem 2.4.

is represented here? Determine the process parameters K_p, L and T with help of the 63-percent method. The time scale is in seconds.

2.5 Figure E-2.3 shows a simulated step response experiment to determine a process model. Which type of process is this, that is, which of the six process types contribute to the process? Calculate the process parameter K_p, L and T by using the 63-percent method. The time scale is in seconds.

2.6 Figure E-2.4 shows a simulated step response experiment to determine the process model. What type of process is this, that is, which of the six process types contributes to the process? Compute the process parameters K_v and L using the method described in Section 2.4 of the book. The time scale is in seconds.

2.7 Figure E-2.5 shows two setpoint changes made to a flow process. Determine the process gain K_p of the two step responses. The measurement range of the flow is between 0 m³/h and 70 m³/h, the controller output is given in percent and the time scale is in seconds.

2.8 Figure E-2.6 shows a setpoint change of a flow process in a paper mill. Determine the process gain K_p from the step response. The flow has a measurement range from 0 to 4 litres/minute, the controller output is given in percent and the time scale is in seconds.

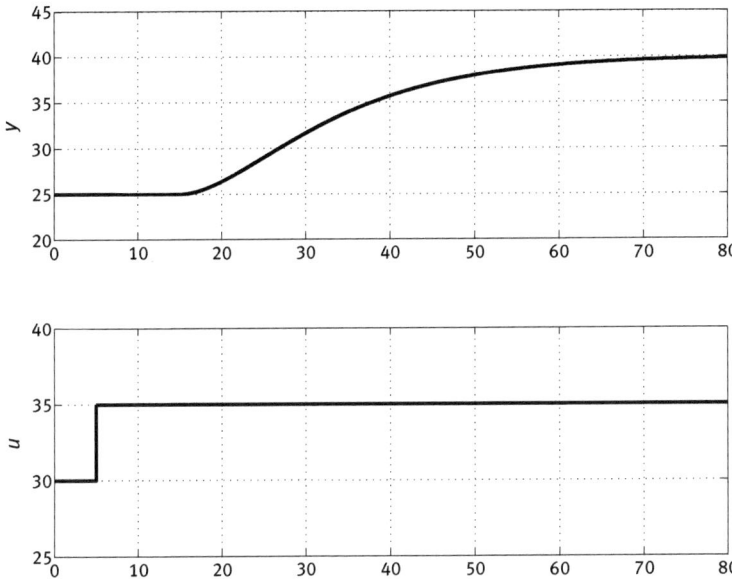

Fig. E-2.3: Step response experiment in Problem 2.5.

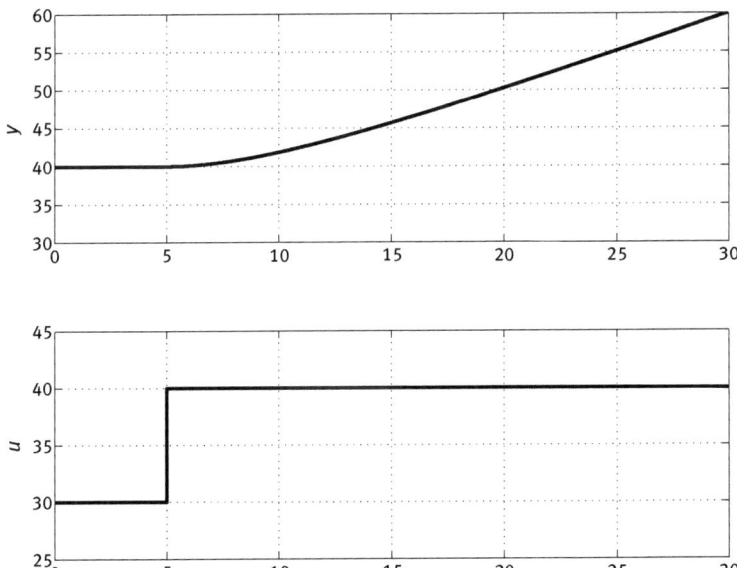

Fig. E-2.4: Step response experiment in Problem 2.6.

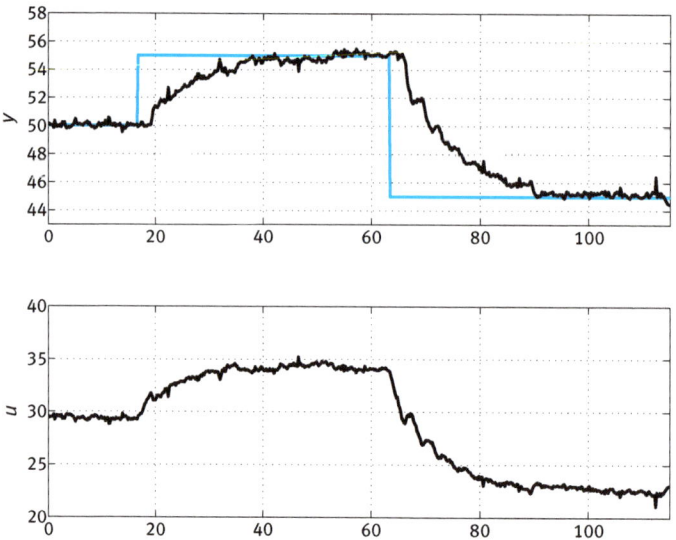

Fig. E-2.5: Setpoint change in the flow process of Problem 2.7.

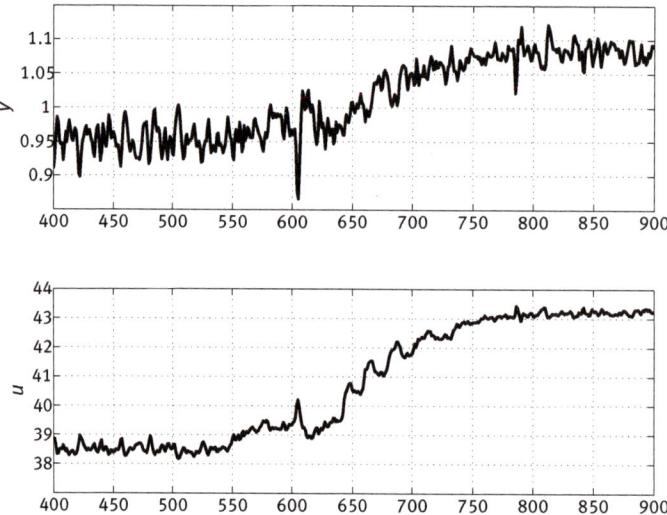

Fig. E-2.6: Setpoint change of a flow loop in Problem 2.8.

2.9 Figure E-2.7 shows a step response experiment of a concentration process in a paper mill. The process variable is a mass concentration and has a measurement range from 2 % to 5 %. The controller output is within a range from 0 % to 100 % and controls a valve that in turn controls the water supply to the mass flow. Thus, a reduction in the

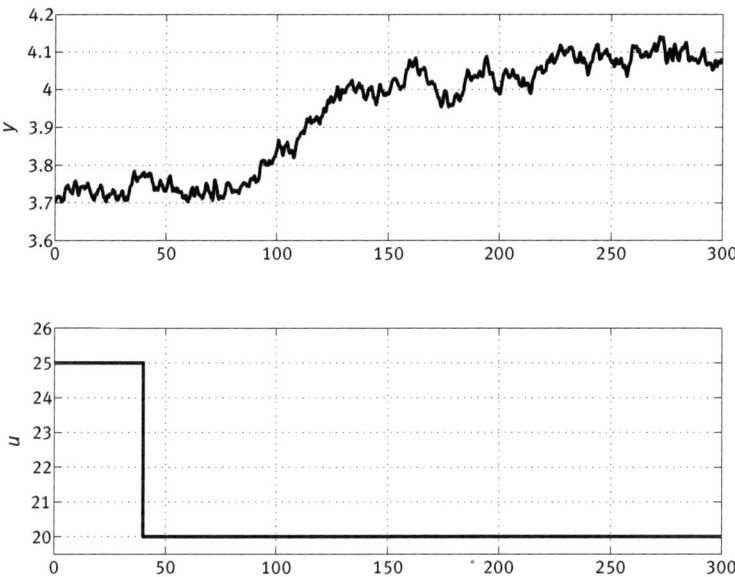

Fig. E-2.7: Step response experiment of a concentration process in Problem 2.9.

controller output gives an increase in the mass concentration. Time scale is in seconds. Determine K_p, L and T of the process using the 63-percent method.

2.10   Figure E-2.8 shows a step response experiment for a concentration process in a paper mill. The process is the same as in Problem 2.9 but with different piping. The process variable is a mass concentration and has a measurement range between 2 % and 5 %. The controller output lies between 0 % and 100 % and controls a valve which in turn controls the water supply to the mass flow. Thus, a reduction in the controller output gives an increase in the mass concentration. Time scale is in seconds. Determine K_p, L and T of the process using the 63-percent method.

2.11   Figure E-2.9 shows two step responses that were derived for a temperature controller in an oven. The time scale is in seconds. The temperature has a measurement range from 0 °C to 900 °C. Determine the two sets of process parameter K_p, L and T for the two step responses using the 63-percent method.

2.12   Figure E-2.10 shows a step response experiment conducted for a flow loop in a paper mill. The step is very small which results in a relatively high noise level. Nevertheless, try to determine K_p, L and T for the process using the 63-percent method. The flow has a measurement range of 0 to 4 litres/minute and the controller signal is given in percent. The time scale is in seconds.

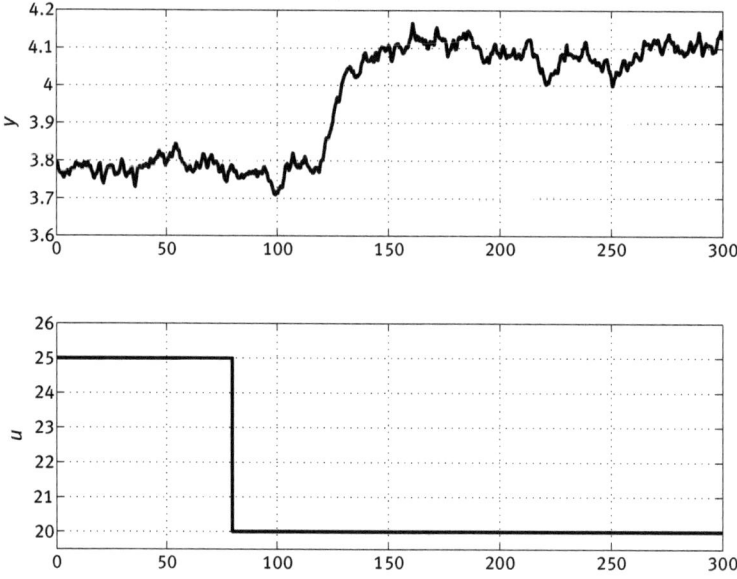

Fig. E-2.8: Step response experiment of the process described in Problem 2.10.

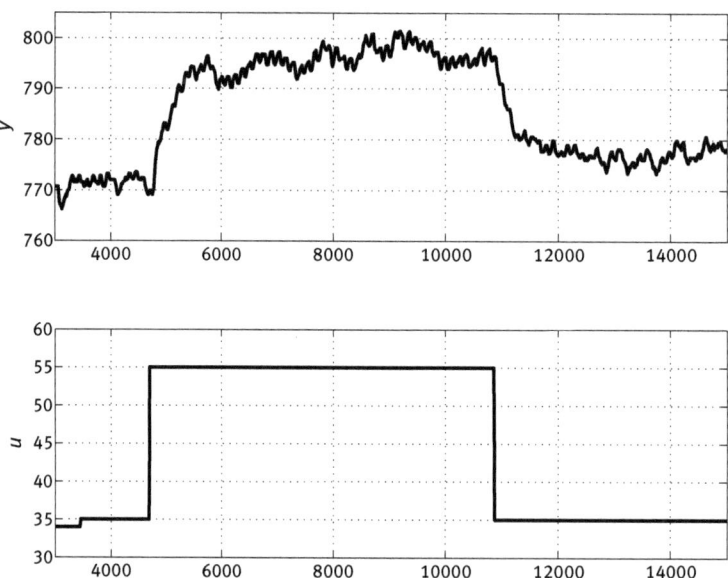

Fig. E-2.9: Step response experiment of the temperature process in Problem 2.11.

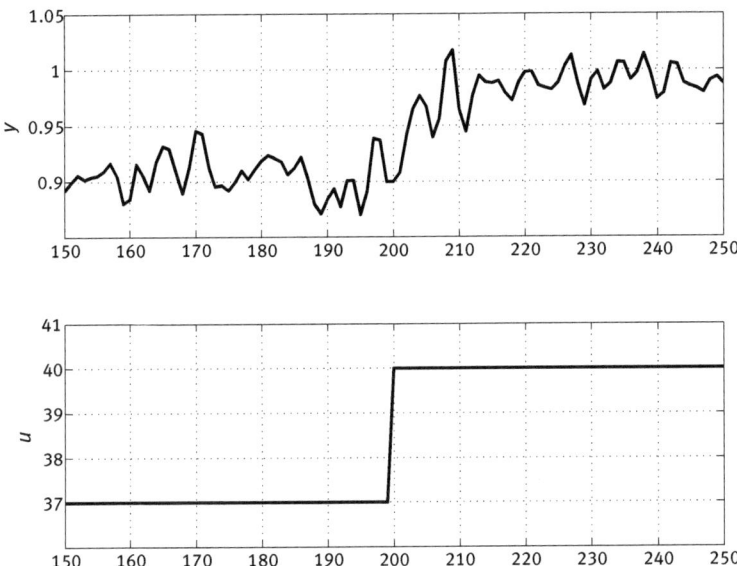

Fig. E-2.10: Step response of the flow control in Problem 2.12.

3. PID-Controllers

3.1 A PI-controller with controller parameters $K = 2$ and $T_i = 30\,\text{s}$ should be replaced with a controller where the proportional band is specified instead of gain and instead of integral time, the number of repetitions per minute is specified. Which parameters should the new controller have?

3.2 A PI-controller with controller parameters $PB = 200\,\%$ and $1/T_i = 0.02$ rps should be replaced with a controller where the gain is specified instead of proportional band and where the integral time is specified in seconds. What will be the parameters in the new controller?

3.3 A PID-controller in series form has the following parameters: $K' = 2$, $T'_i = 20\,\text{s}$ and $T'_d = 4\,\text{s}$. The controller should be replaced with a new controller in parallel form. What are the controller parameters of the new controller?

3.4 A PID-controller in parallel form has the parameters $K = 3$, $T_i = 30\,\text{s}$ and $T_d = 4.8\,\text{s}$. The controller should be replaced with a new controller in series form. What are the controller parameters of the new controller?

3.5 A common rule of thumb for choosing the derivative time in a PID-controller is to choose it to be one quarter of the integral time. Assume that you have a controller in series form and follow this rule of thumb, that is, select $T'_d = T'_i/4$. If this controller is changed to a controller in parallel form and one calculates the new controller parameters according to the conversion formulas

$$T_i = T'_i + T'_d$$

$$T_d = \frac{T'_i T'_d}{T'_i + T'_d},$$

what will be the new ratio T_d/T_i between the derivative time and the integral time?

3.6 Assume that we use the same rule of thumb as in Problem 3.5, but that we now start with a controller in parallel form and choose $T_d = T_i/4$. If we switch to a controller in series form with the conversion equations

$$T'_i = \frac{T_i}{2}\left(1 + \sqrt{1 - \frac{4T_d}{T_i}}\right)$$

$$T'_d = \frac{T_i}{2}\left(1 - \sqrt{1 - \frac{4T_d}{T_i}}\right)$$

what will the ratio T'_d/T'_i between derivative and integral time be?

3.7 A person has tuned a PID-controller by hand. The goal was to get a response that is as fast as possible when dealing with setpoint changes, but without an overshoot. After doing this it was found that the response to load disturbances was too slow. Is there any way to solve the problem, i.e. to get good control both in case of setpoint changes and load disturbances?

3.8 A process is controlled by a PID-controller in the form

$$u = K\left(br - y + \frac{1}{T_i}\int e(t)dt - T_d\frac{dy}{dt}\right)$$

The original controller parameters are $K = 1.7$, $T_i = 1.3$, $T_d = 0.4$ and $b = 1$. Figure E-3.1a shows the response to setpoint changes with this controller. The remaining subplots, Figure E-3.1b – Figure E-3.1d, show the control that results after the following changes have been made:
1) Remove the D-part, that means applying a PI-controller.
2) Remove the I-part and the D-part, that means applying a P-controller.
3) Change the setpoint weight to $b = 0$.
Match the three changes with the three subplots E-3.1b – E-3.1d.

3.9 You want to develop a model of a temperature process, but lack information about the process variable range PV_{range} for the temperature. Therefore, the following experiment is carried out. The controller is first changed to be a P-controller with

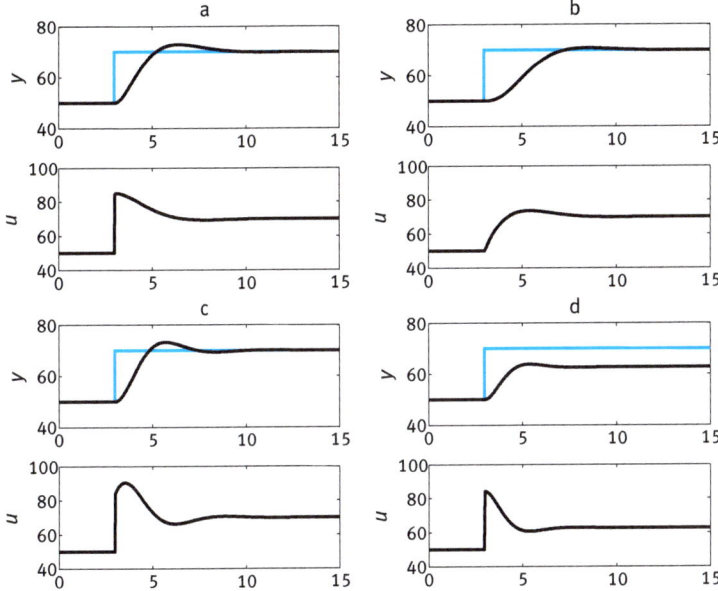

Fig. E-3.1: Setpoint changes for different controllers in Problem 3.8.

controller gain $K = 0.5$. When the control loop is in steady-state, a step change is made to the setpoint with the size $100\,°C$. This results in the controller output momentarily changing by $10\,\%$. The controller output range is as usual $OP_{range} = 100\,\%$. Use this information to determine the measurement range of the temperature.

3.10 The information about the process variable range for a flow control loop is missing. To determine this, you do the experiment shown in Figure E-3.2. The controller is a P-controller with gain $K = 1.5$. When the control loop is in steady-state, a step change is made to the setpoint. The controller output has the range $OP_{range} = 100\,\%$ and the flow is shown in m^3/h. Determine the measurement range PV_{range}.

4. PID-Controller Tuning

4.1 Figure E-4.1 shows two different responses to load disturbances of a control loop with poorly tuned PI-controllers. One response corresponds to a badly tuned gain, and the other to a badly tuned integral time. How should the controller settings be changed in the two cases?

4.2 For a certain process, a PI-controller has been tuned with parameters $K = 2$ and $T_i = 5$ s. The response to a setpoint change is shown in Figure E-4.2a. The other sub-

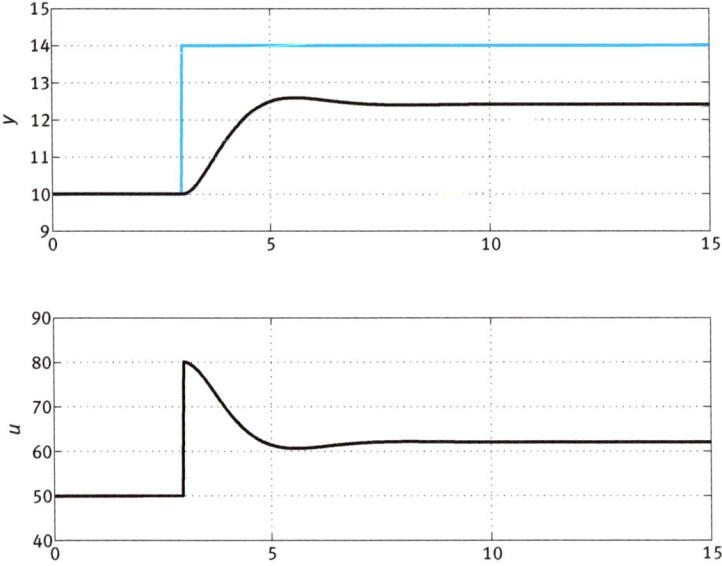

Fig. E-3.2: Experiment to determine the measurement range in Problem 3.10.

plots in Figure E-4.2 show responses to setpoint changes with the following controller parameters:

1) $K = 6$, $T_i = 5$ s.
2) $K = 2$, $T_i = 15$ s.
3) $K = 6$, $T_i = 15$ s.

Match the three controller tuning changes with the three subplots E-4.2b – E-4.2d.

4.3 Apply the process parameters K_p, L and T from the process in Problem 2.4 to calculate the controller parameters of a PI-controller using the AMIGO method, Lambda

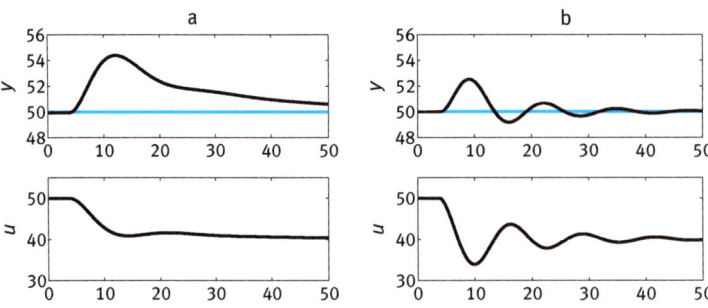

Fig. E-4.1: Load disturbances in Problem 4.1.

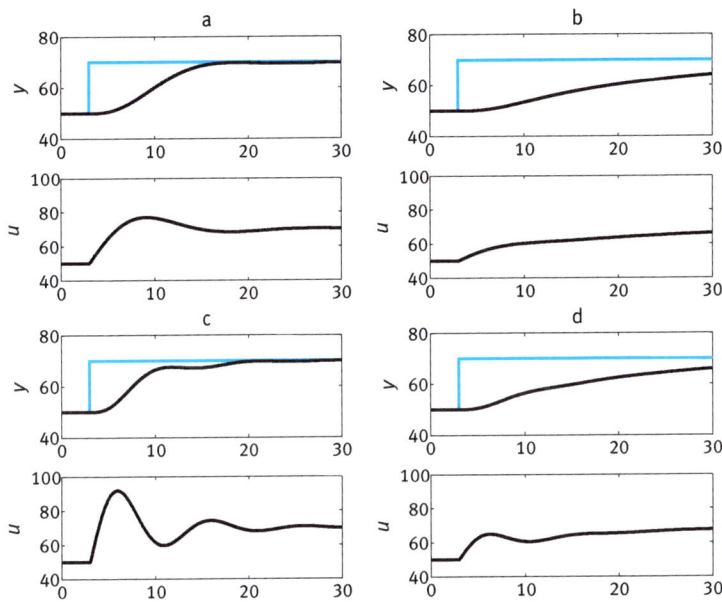

Fig. E-4.2: Setpoint changes in Problem 4.2.

method with $\lambda = T$ and the One-third rule. For the One-third rule you can use the relationship $T_p \approx L + 3T$. Compare the different settings.

4.4 Apply the process parameters K_p, L and T obtained for the process in Problem 2.5 to calculate the controller parameters for a PI- and a PID-controller in parallel form using the AMIGO and Lambda method with $\lambda = T$. Compare the tuning settings.

4.5 Use the process parameters K_v, and L obtained for the integrating process in Problem 2.6 to calculate the controller parameters for PI- and PID-controllers in parallel form using the Ziegler-Nichols method and the AMIGO method. Compare the settings.

4.6 Apply the process parameters K_p, L and T obtained for the process in Problem 2.9 to calculate the controller parameters for PI- and PID-controllers in parallel form using the AMIGO method and the Lambda method with $\lambda = T$. Compare the tuning settings.

4.7 Use the process parameters derived for the process in Problem 2.10 to calculate the controller parameters for a PI-controller using the AMIGO method, the Lambda method, and the One-third rule. Since this is a process with long dead time, we choose λ in the Lambda method to $\lambda = 3L$. Determine the process time T_p by using $T_p \approx L+3T$, and by examining Figure E-2.8. Compare the tuning settings.

4.8 Two step responses from a temperature process were analyzed in Problem 2.11. Determine a set of controller parameters by averaging the parameters from the two experiments. In this case, the dead time is $L = 0$. Which of the four tuning methods for PI-controllers shown in Table 4.7 can be used for this type of process? Calculate PI-controllers for the methods that can be used and compare the results.

4.9 Use the process parameter K_p, L and T resulting from Problem 2.12 to calculate the controller parameters of a PI-controller using the Ziegler-Nichols method, AMIGO method, Lambda method with $\lambda = T$ and the One-third rule. Choose $L = 2$ and determine the process time by using the equation $T_p = L + 3T$. Compare the controller settings.

4.10 If a process has a very long dead time L compared to the time constant T, one can neglect the time constant in the tuning rules, i.e. choose $T = 0$. Investigate which controller parameter the AMIGO method and the Lambda method with $\lambda = 3L$ provide for PI- and PID-controllers and compare the results.

4.11 How does the One-third rule work for the processes with very long dead time that was analysed in Problem 4.10? Consider what the step response looks like and calculate the controller parameters using the One-third rule.

4.12 A self-oscillation experiment for a process gave the process parameters $K_c = 6$ and $T_c = 20\,\mathrm{s}$. A step response experiment showed that the process gain was $K_p = 2$. Use these process parameters to determine the parameters of a PI-controller using the Ziegler-Nichols method and the AMIGO method. Compare the results.

5. Nonlinear Processes

5.1 Figure E-5.1 shows a series of step responses of a simulated process with a non-linear valve. The controller output is changed in 10 % steps. Sketch the valve characteristic using the figure.

5.2 Figure E-5.2 shows the valve characteristic of a nonlinear valve that will be used to control a flow. You want to apply gain scheduling with three different controller parameter settings to compensate for the nonlinearity. Thus, you need to define two limits: the first between setting 1 and 2 and the second between setting 2 and 3. Suggest a good choice of a reference variable GS_{ref} and good choices for the values of the two limits.

5.3 To determine the size of the backlash in a valve, step changes have been made in the controller output as shown in Figure E-5.3. Use the result to determine the size

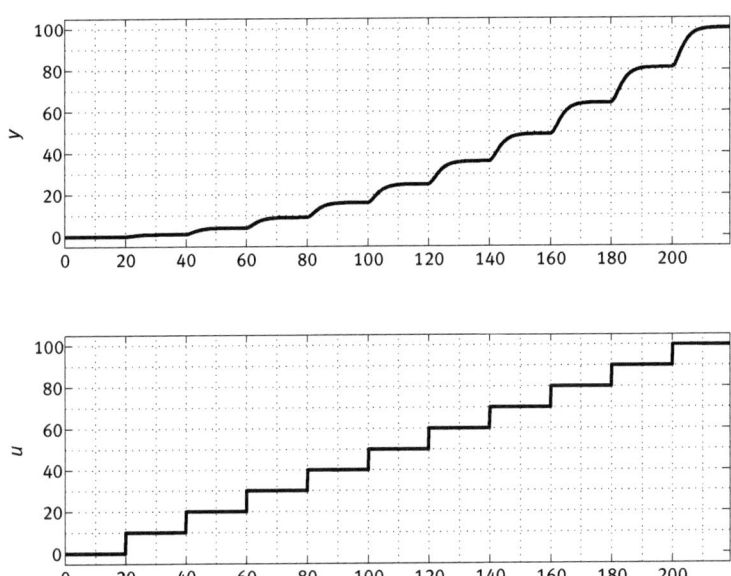

Fig. E-5.1: Step responses for determining the valve characteristic in Problem 5.1.

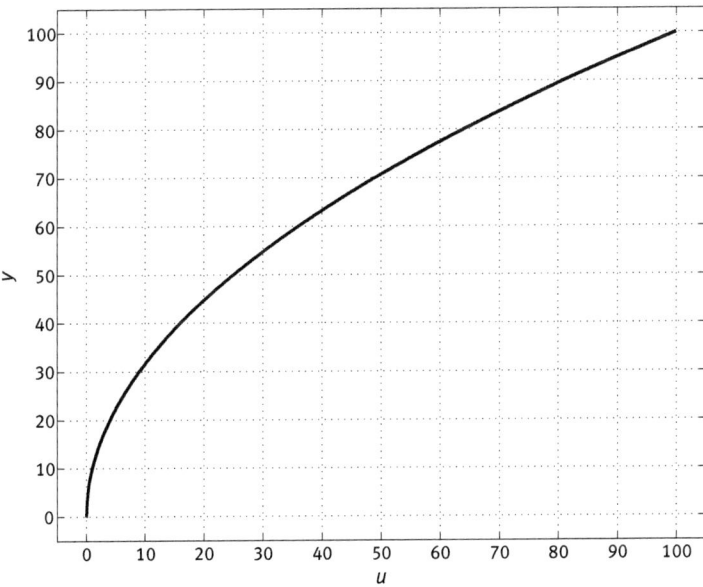

Fig. E-5.2: Valve characteristic for the valve used in Problem 5.2.

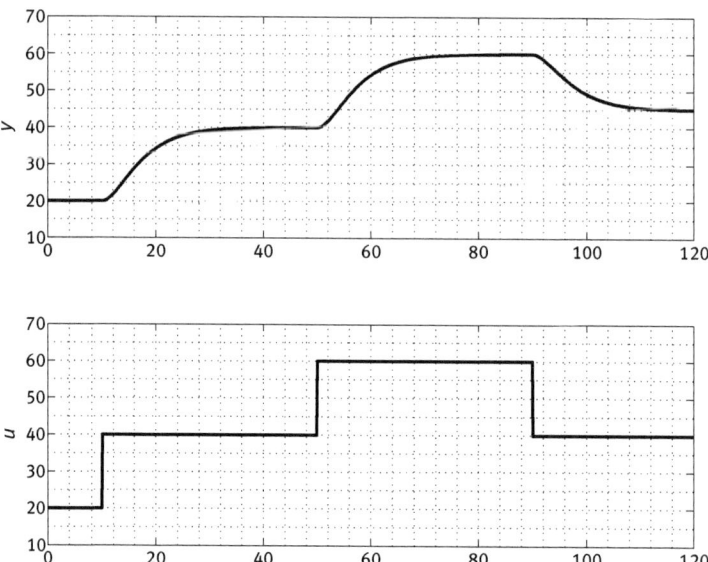

Fig. E-5.3: Step response experiment to determine backlash in Problem 5.3.

of the backlash gap Δu. The process has the process gain $K_p = 1$. Can we see the gain in the figure?

5.4 Figure E-5.4 shows the signals from a flow control loop in a paper mill. The flow here is the return pulp to a paper machine, the process variable range is 0 to 10800 l/min and the time scale is seconds. The control loop oscillates. Explain why.

5.5 Figure E-5.5 shows a setpoint change in a flow control loop with a PI-controller. The control is slow and the response is oscillating. Explain the reason for the poor control performance.

5.6 Figure E-5.6 shows step response experiments in a flow loop to analyse the extent of backlash in the control valve. Apply the response in Figure E-5.6 to determine the backlash. Is the valve characteristic quick opening, linear or equal percentage?

5.7 Figure E-5.7 shows a three hour long recording of data from a concentration process in a paper mill. The concentration fluctuates close to the setpoint but the controller output varies rather a lot. What can this mean?

5.8 Figure E-5.8 shows the process variable and controller output of a setpoint change in a concentration control loop. The process variable looks fine but the controller output drifts slowly after 200 seconds. What could be the reasons for this?

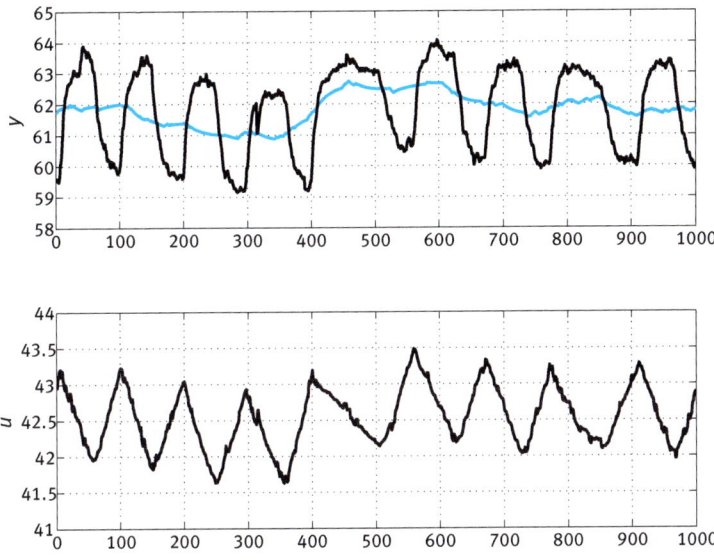

Fig. E-5.4: Oscillating control loop in Problem 5.4. The cyan line in the upper plot is the setpoint.

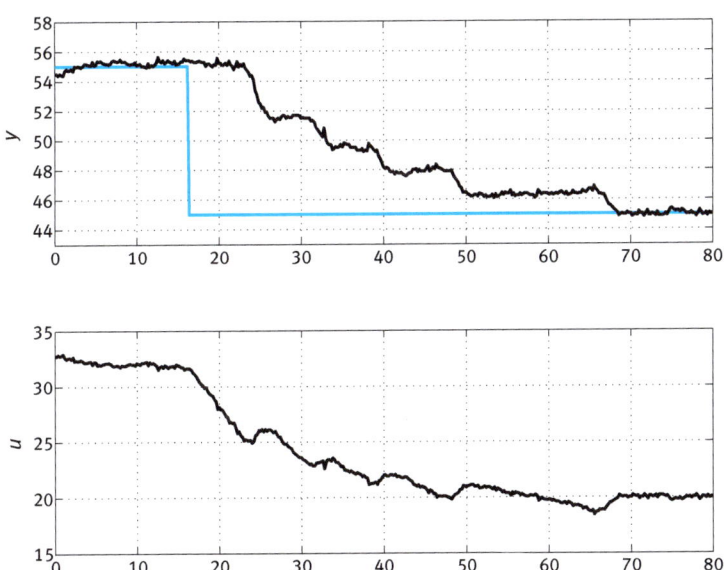

Fig. E-5.5: Setpoint change in Problem 5.5.

5.9 Figure E-5.9 shows the result from studying a valve in a concentration control loop. Small successive step changes in the controller output were made to determine the size of the backlash. How big is the gap of the backlash?

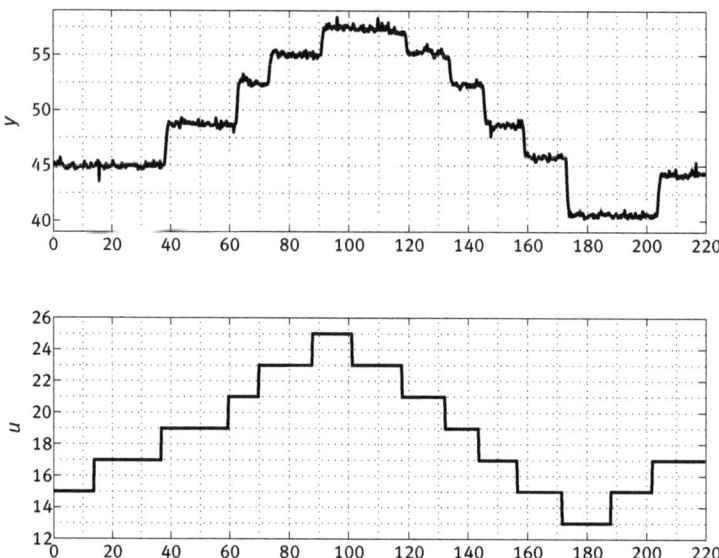

Fig. E-5.6: Step response experiments in Problem 5.6.

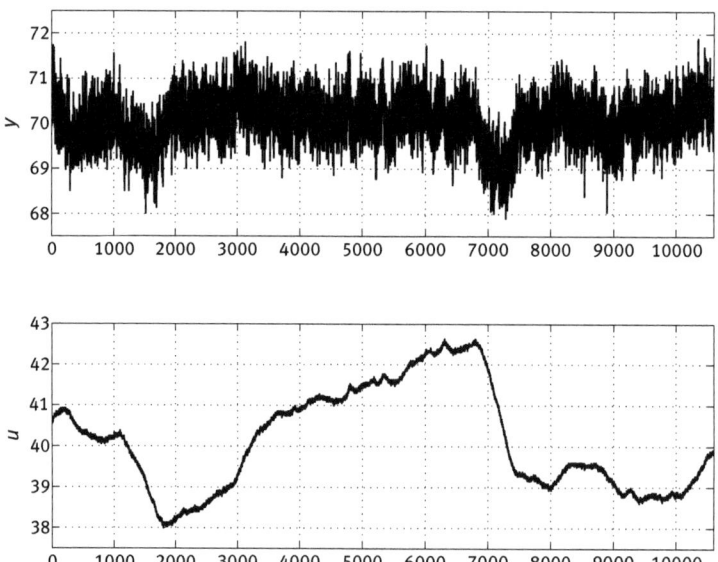

Fig. E-5.7: Concentration control loop in Problem 5.7. Time scale is given in seconds.

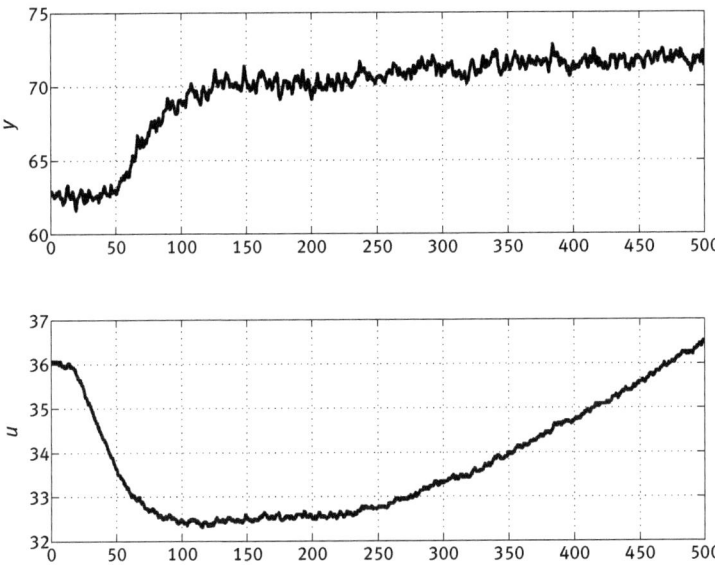

Fig. E-5.8: Concentration control loop in Problem 5.8. Time scale is in seconds.

Fig. E-5.9: Valve examination in Problem 5.9.

6. Control Strategies

6.1 Figure E-6.1a shows a signal that consists of the sum of two sine waves, one with frequency 0.1 rad/s and an oscillation period of 63 s and one with frequency 5 rad/s and an oscillation period of 1.3 s. The other subplots b – f show the corresponding signals filtered with three low pass filters with time constants $T_{lp} = 0.5$ s, $T_{lp} = 1$ s and $T_{lp} = 20$ s and two high pass filters with time constants $T_{hp} = 0.1$ s and $T_{hp} = 1$ s. Match the five subplots with the corresponding five filters.

6.2 In a control loop with controller parameters $K = 0.5$ and $T_i = 30$ s there is a measurement disturbance in the form of a sine wave with a period time of 1 s. To reduce the impact of the disturbance, the process variable is sent through a low-pass filter. Suggest a suitable time constant for the filter.

6.3 In a flow control loop, someone has placed a transducer with a low-pass filter that has a time constant of 30 seconds. One notices that the sensor makes the flow control too slow and therefore wants to make the process variable react more quickly to variations in the flow by introducing a lead/lag filter. How should the filter parameters be chosen if you want the measured value to react as if the sensor had a filter with a time constant of 5 seconds instead of 30 seconds?

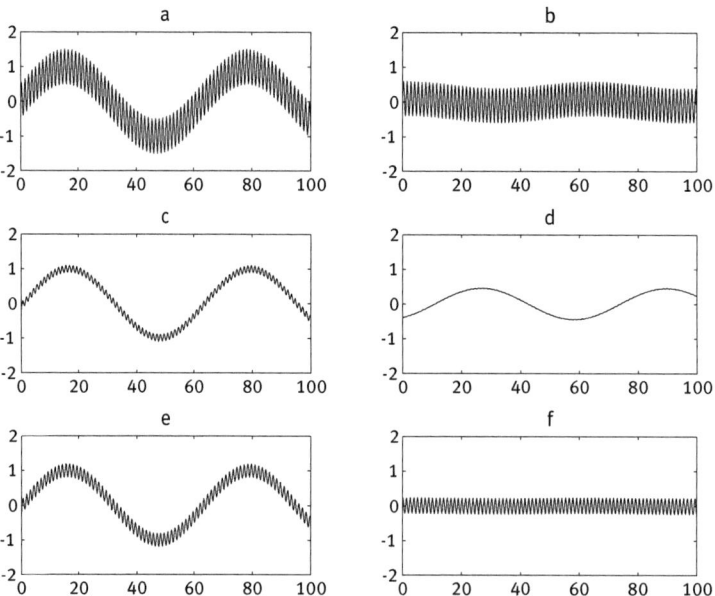

Fig. E-6.1: Signals in subplot a are filtered with different filters in subplots b – f as described in Problem 6.1.

6.4   Figure E-6.2 shows a scheme consisting of three selectors, A, B and C, which can be either MAX selector or MIN selector. Selector A has the signals 10 % and 20 % as input signals, selector B has the signals 30 % and 40 % as input signals and selector C has the output signals from A and B as input signals and the signal y as output.

a)   What is the value of y if all selectors are MAX-selectors?

b)   What is the value of y if all selectors are MIN-selectors?

c)   Give selectors that result in a value of $y = 20\,\%$.

d)   Give selectors that result in a value of $y = 30\,\%$.

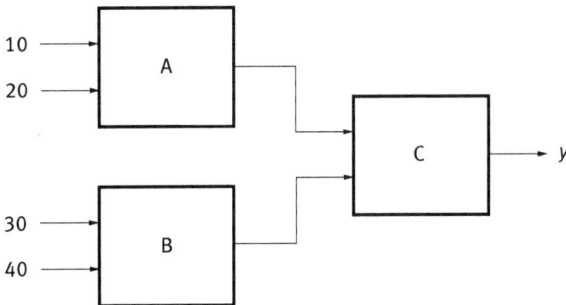

Fig. E-6.2: Schema with selectors in Problem 6.4.

6.5   In a control loop, there is a load disturbance that is measurable and that causes problems in keeping the process variable at its setpoint. Therefore, it is decided to introduce feed forward of the disturbance signal. You want to use static feed forward and to determine an appropriate value of K_f you did the following experiment. First, the values of the controller output and the disturbance signal were noted at a time when the control worked well and the process variable was at the setpoint. These were $u_1 = 30\,\%$ and $v_1 = 45\,\%$. Later, when the disturbance signal had been changed to a new value, $v_2 = 50\,\%$, and the control worked well, it was noted that the controller output was changed to $u_2 = 20\,\%$. Use this information to determine K_f.

6.6   Figure E-6.3 shows a simulated experiment of a control loop with a measurable load disturbance v. You want to include the load disturbance in a feed forward structure to the controller for better controlling the load changes. Use the figure to determine an appropriate feed forward gain.

6.7   Figure E-6.4 shows the block diagram for cascade control of a heat exchanger. It has been discovered that there are large variations in the temperature of the incoming water on the secondary side. A suggestion is therefore to install a sensor for this

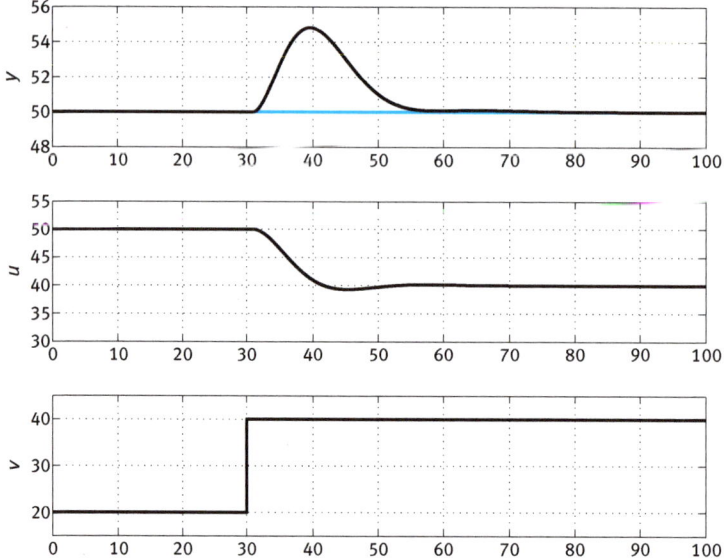

Fig. E-6.3: Experiment with load disturbance in Problem 6.6.

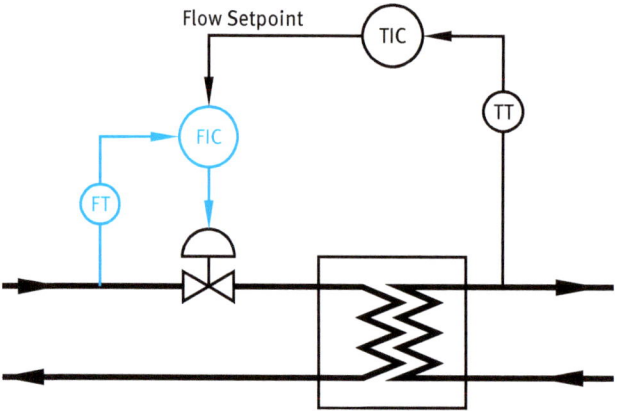

Fig. E-6.4: Cascade control of a heat exchanger in Problem 6.7.

temperature and use the signal as a feedforward signal. In order to obtain a compensation that is as fast as possible, someone suggests that the temperature signal should be connected to the flow controller FIC. Is this a good suggestion? Think about how the control will work in the event of a sudden change in the input temperature. How should the feed-forward be set up?

6.8 For the control of a boiler you have the option of raising the temperature with two different burners, one that is powered by excess gas that you have free access to and one that is powered by natural gas that you can buy. It is therefore desired that the burner with excess gas should work as much as possible, while the natural gas is only used when the excess gas is not enough. Suggest a suitable method to solve this control problem.

Solutions to the Exercises

2. Process Types and Step Response Analysis

2.1   a) First order with dead time. b) Integrating and oscillating.
c) Higher order with dead time. d) Integrating and first order.
e) Oscillating with dead time. f) Inverse response with dead time.

2.2   The change in the controller output is given by $\Delta u = 30 - 20 = 10\%$ and the change in the process variable is given by $\Delta y = 60 - 40 = 20\%$. Since both the controller output and the process variable have the measurement range 0 to 100 percent, $PV_{range} = OP_{range} = 100\%$ and the process gain is given by

$$K_p = \frac{\Delta y/PV_{range}}{\Delta u/OP_{range}} = \frac{20}{10} = 2$$

2.3   The change in the controller output signal is $\Delta u = 40 - 30 = 10\%$ and the range of the controller output is $OP_{range} = 100\%$. The change in the process variable was noted as $\Delta y = 85 - 70 = 15\,°C$ and the process variable range was $PV_{range} = 200 - 50 = 150\,°C$. These parameters give a process gain of

$$K_p = \frac{\Delta y/PV_{range}}{\Delta u/OP_{range}} = \frac{15/150}{10/100} = 1$$

2.4   This is a first order process with dead time. Figure S-2.1 shows how to determine the process parameters. The step changes in the process variable and the controller output are $\Delta y = 55 - 35 = 20\%$ and $\Delta u = 40 - 30 = 10\%$. Since $PV_{range} = OP_{range} = 100\%$ this means that $K_p = \Delta y/\Delta u = 2$. The step change in the controller output takes place at time $t = 5$ s. The intersection between the increase of the process variable and the level of the process variable before the step change takes place at time $t = 10$ s. This gives $L = 10 - 5 = 5$ s. The process variable has reached 63 % of its final value when it has reached up to $y = 35 + 0.63\Delta y = 47.6\%$. This happens at time $t = 25$ s. Since the dead time has ended at time $t = 10$ s this results in a time constant $T = 25 - 10 = 15$ s.

2.5   This is a higher order process with dead time. Figure S-2.2 shows how to calculate the process parameters. The steady state change in the process variable and controller output is $\Delta y = 40 - 25 = 15\%$ and $\Delta u = 35 - 30 = 5\%$. Because $PV_{range} = OP_{range} = 100\%$ it follows that $K_p = \Delta y/\Delta u = 3$. The step response change in the controller output occurs at time $t = 5$ s. The intersection between the tangent of the process variable and the steady state of the process variable before the step change is at time $t = 18$ s. This results in a dead time of $L = 18 - 5 = 13$ s. The process variable has

https://doi.org/10.1515/9783111104959-008

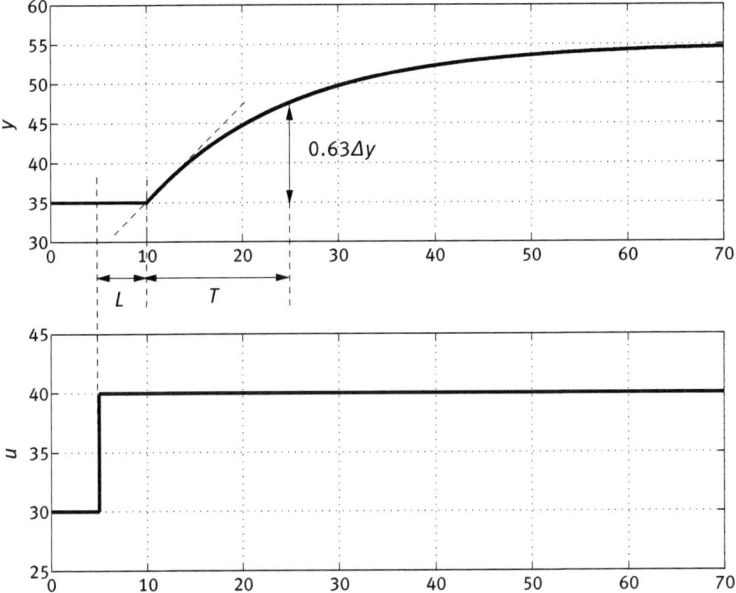

Fig. S-2.1: Step response experiment in Problem 2.4.

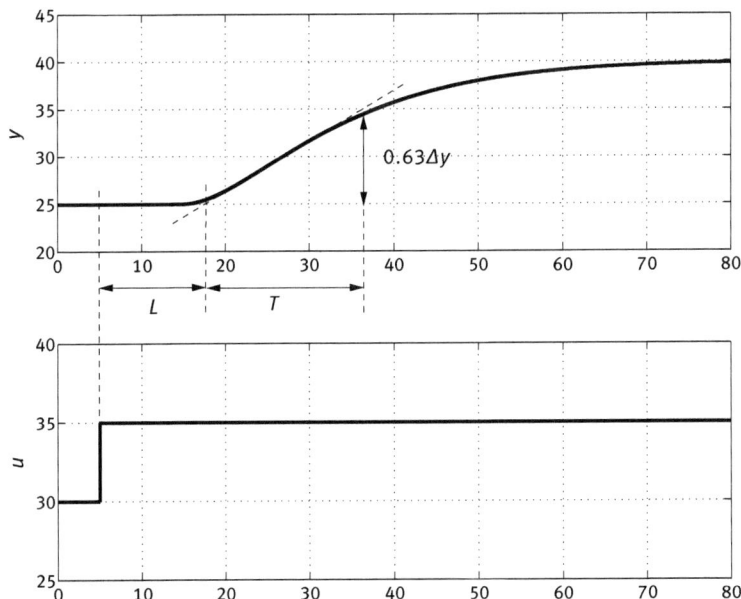

Fig. S-2.2: Step response experiment in Problem 2.5.

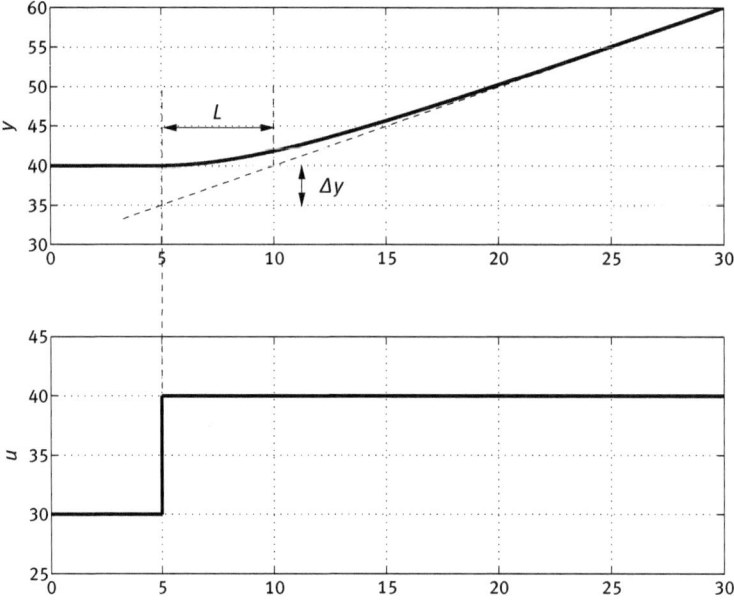

Fig. S-2.3: Step response experiment in Problem 2.6.

reached 63 % of its final value when it has reached $y = 25 + 0.63\Delta y = 34.5\,\%$. This occurs at time $t = 37$ s. Since the dead time has elapsed at time $t = 18$ s this gives a time constant of $T = 37 - 18 = 19$ s.

2.6 This is an integrating and first order process. Figure S-2.3 shows how to determine the process parameters. The step change $\Delta u = 40 - 30 = 10\,\%$ occurs at time $t = 5$ s. The value of the process variable before the step change was made is $y = 40\,\%$. The intersection of the slope in the process variable and this value occurs at time $t = 10$ s. This means that the dead time can be calculated as $L = 10 - 5 = 5$ s. If we extend the slope of the process variable to the time of the step change, $t = 5$ s, then we can note the value which would have occurred at this time, which is $y = 35\,\%$. This means that the process variable change is $\Delta y = 40 - 35 = 5\,\%$. Since $OP_{\text{range}} = PV_{\text{range}} = 100\,\%$, we can determine the speed gain as described in Section 2.4 as

$$K_v = \frac{\Delta y/PV_{\text{range}}}{\Delta u/OP_{\text{range}} \cdot L} = \frac{5}{10 \cdot 5} = 0.1\,\text{s}^{-1}$$

2.7 The flow changes from $50\,\text{m}^3/\text{h}$ to $55\,\text{m}^3/\text{h}$ with the first setpoint change. The controller output changes from $29\,\%$ to $34\,\%$. With the second setpoint change the flow decreases from $55\,\text{m}^3/\text{h}$ to $45\,\text{m}^3/\text{h}$. The controller output decreases from $34\,\%$ to $22\,\%$. As the control output range is given as $OP_{\text{range}} = 100\,\%$ and the process variable range $PV_{range} = 70\,\text{m}^3/\text{h}$ we can carry out the two following calculations to determine the the process gain:

$$K_{p1} = \frac{\Delta y_1/PV_{range}}{\Delta u_1/OP_{range}} = \frac{(55-50)/70}{(34-29)/100} \approx 1.4$$

$$K_{p2} = \frac{\Delta y_2/PV_{range}}{\Delta u_2/OP_{range}} = \frac{(45-55)/70}{(22-34)/100} \approx 1.2$$

2.8 The controller output changes from 38.5% to 43.2%, which gives a change of $\Delta u = 43.2 - 38.5 = 4.7\%$. The flow changes from 0.95 l/min to 1.08 l/min. Thus, the process variable change is $\Delta y = 0.13$ l/min. As the controller output range is $OP_{range} = 100\%$ and the process variable range is $PV_{range} = 4$ l/min we get the following estimate of the process gain:

$$K_p = \frac{\Delta y/PV_{range}}{\Delta u/OP_{range}} = \frac{0.13/4}{4.7/100} \approx 0.69$$

2.9 Figure S-2.4 shows how you can derive the process parameters. The controller output changes from 25% to 20%. Therefore, the change is $\Delta u = -5\%$. The controller output range is $OP_{range} = 100\%$. The process variable changes from 3.73% to 4.07%, and therefore the change is $\Delta y = 0.34\%$. The process variable range is $PV_{range} = 5-2 = 3\%$. This gives us a process gain of

$$K_p = \frac{\Delta y/PV_{range}}{\Delta u/OP_{range}} = \frac{0.34/3}{-5/100} \approx -2.3$$

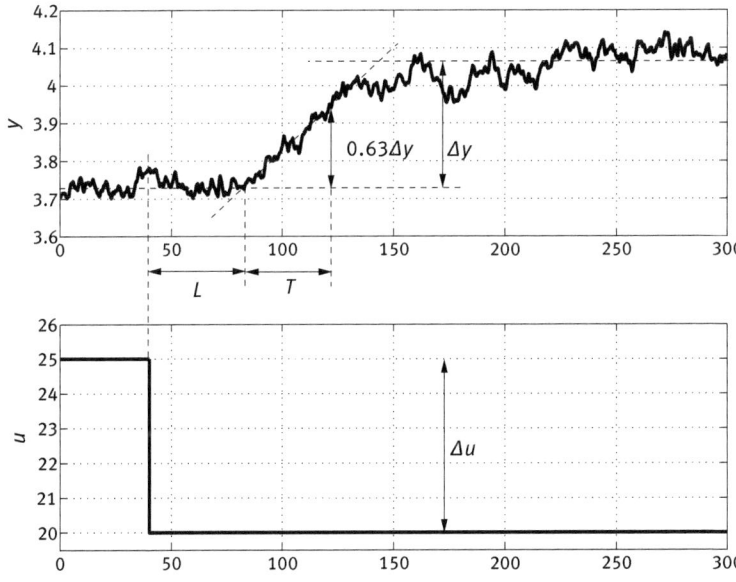

Fig. S-2.4: Step response experiment of the concentration process in Problem 2.9.

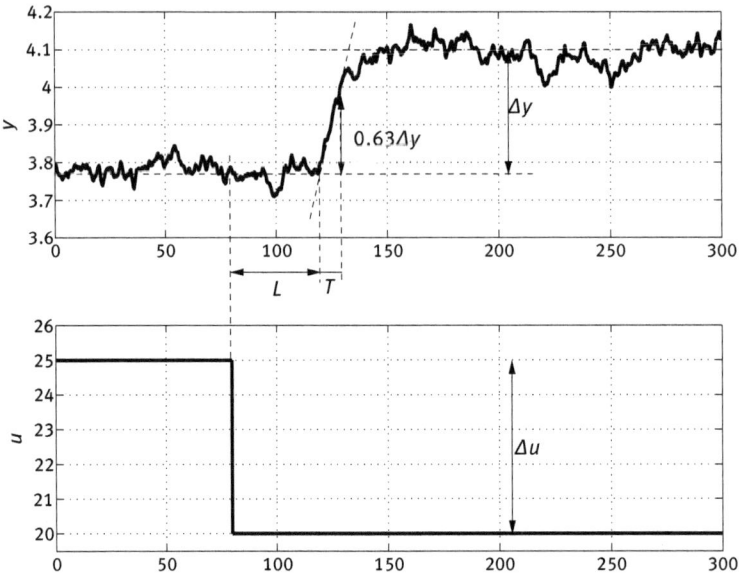

Fig. S-2.5: Step response experiment of the process described in Problem 2.10.

The time of the step change is at $t = 40\,\text{s}$. The tangent line at the steepest slope of the process variable intersects with the initial level of the process variable before the step change at time $t = 85$ s. Thus, the dead time can be calculated as $L = 85 - 40 = 45$ s. The process variable has reached 63 % of the final value when it is at $y = 3.73 + 0.63\Delta y = 3.94\,\%$. This happens at time $t = 122$ s. This gives the time constant as $T = 122 - 85 = 37$ s.

2.10 Figure S-2.5 shows how to derive the process parameter. The controller output changes from 25 % to 20 %, that is, $\Delta u = -5\,\%$. The controller output range is $OP_{\text{range}} = 100\,\%$. The process variable changes from 3.77 % to 4.1 %, that is, $\Delta y = 0.33\,\%$ with a process variable range of $PV_{\text{range}} = 5 - 2 = 3\,\%$. This gives a process gain of

$$K_p = \frac{\Delta y/PV_{\text{range}}}{\Delta u/OP_{\text{range}}} = \frac{0.33/3}{-5/100} \approx -2.2$$

The step changes at time $t = 80$ s. The intersection between the tangent line and the initial constant level of the process variable is at time $t = 120$ s. It follows that the dead time is $L = 120 - 80 = 40$ s. The process variable reaches 63 % when it reaches the level of $y = 3.77 + 0.63\Delta y = 3.98\,\%$. This happens at time $t = 130$ s. Thus, the time constant is $T = 130 - 120 = 10$ s.

2.11 Figure S-2.6 gives some reference lines. It is not so easy to measure the parameters accurately in the figure but this is often the case when working with industrial

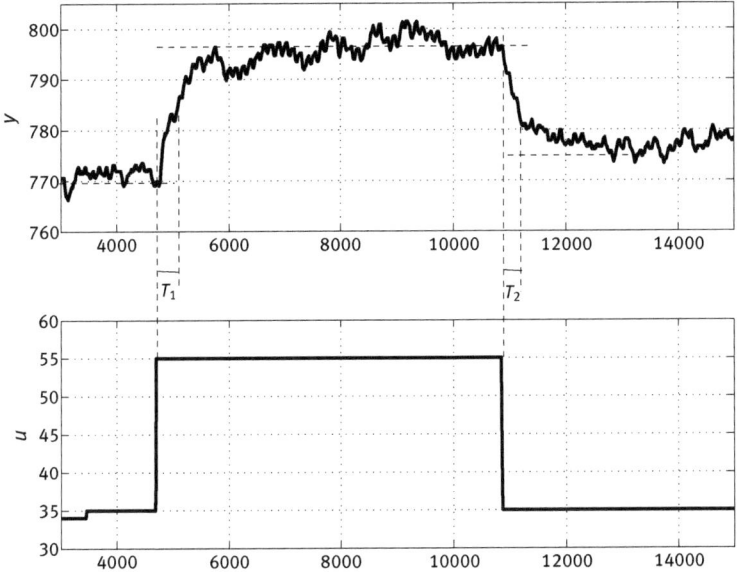

Fig. S-2.6: Step response experiment of Problem 2.11.

data. Both steps are of magnitude $\Delta u = 20\%$, but in differnt directions and thus with different signs. The three stationary levels of the process variables are difficult to determine. We estimate that the process variable starts at $770\,°C$, then lies around a level of $796\,°C$ and finally goes back to $775\,°C$. The two changes in y are therefore $\Delta y_1 = 796 - 770 = 26\,°C$ and $\Delta y_2 = 775 - 796 = -21\,°C$. Since the process variable range is $PV_{\text{range}} = 900\,°C$ and the controller output range is $OP_{\text{range}} = 100\%$ we can calculate the two process gains as

$$K_{p1} = \frac{\Delta y_1/PV_{\text{range}}}{\Delta u_1/OP_{\text{range}}} = \frac{26/900}{20/100} \approx 0.14$$

$$K_{p2} = \frac{\Delta y_2/PV_{\text{range}}}{\Delta u_2/OP_{\text{range}}} = \frac{-21/900}{-20/100} \approx 0.12$$

The first step is at time $t = 4700\,s$ and the second at time $t = 10900\,s$. The vertical lines that we draw from the step changes show that the dead time is negligible, in other words we can set $L = 0$. The process variable has reached 63 % of the final value when $y = 770+0.63·26 = 786\,°C$ and $y = 796-0.63·21 = 783\,°C$ respectively. Vertical lines in the figure indicate the time when the step response reaches those values. This is approximately at $t = 5100\,s$ and $t = 11200\,s$. This gives us the two time constants as $T_1 = 5100 - 4700 = 400\,s$ and $T_2 = 11200 - 10900 = 300\,s$.

2.12 This is a step response with a high noise level and it is also sampled rather sparsely, which contributes further to the uncertainty. You have to try to ignore the noise and see the actual underlying flow. A line representing the estimated flow is

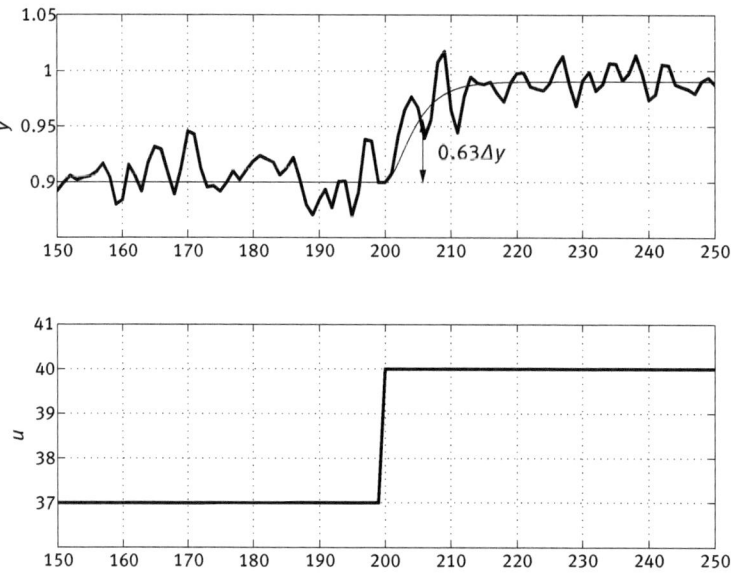

Fig. S-2.7: Step response experiment of a flow process in Problem 2.12.

drawn in Figure S-2.7. The controller output is $\Delta u = 40 - 37 = 3$ %. The flow changes from 0.9 l/min to 0.99 l/min. This means that $\Delta y = 0.09$ l/min. Since controller output range is $OP_{range} = 100$ % and the process variable range is $PV_{range} = 4$ l/min we get the following process gain:

$$K_p = \frac{\Delta y / PV_{range}}{\Delta u / OP_{range}} = \frac{0.09/4}{3/100} \approx 0.75$$

It is difficult to estimate the dead time L of the process. It is something between 0 and 2 seconds. The flow has reached 63 % of its new level when it is $y = 0.9 + 0.63\Delta y = 0.96$ l/min. This occurs approximately at time $t = 206$ s which gives us the time constant $T = 206 - 200 = 6$ s.

3. PID-Controllers

3.1

$$PB = \frac{100}{K} = \frac{100}{2} = 50 \%$$

$$\frac{1}{T_i} = \frac{1}{30} \text{ rps} = \frac{60}{30} \text{ rpm} = 2 \text{ rpm}$$

3.2

$$K = \frac{100}{PB} = \frac{100}{200} = 0.5$$

$$T_i = \frac{1}{0.02} s = 50\,s$$

3.3

$$K = K' \frac{T'_i + T'_d}{T'_i} = 2\frac{20 + 4}{20} = 2.4$$

$$T_i = T'_i + T'_d = 20 + 4 = 24\,s$$

$$T_d = \frac{T'_i T'_d}{T'_i + T'_d} = \frac{20 \cdot 4}{20 + 4} \approx 3.33\,s$$

3.4

$$K' = \frac{K}{2}\left(1 + \sqrt{1 - \frac{4T_d}{T_i}}\right) = \frac{3}{2}\left(1 + \sqrt{1 - \frac{4 \cdot 4.8}{30}}\right) = 2.4$$

$$T'_i = \frac{T_i}{2}\left(1 + \sqrt{1 - \frac{4T_d}{T_i}}\right) = \frac{30}{2}\left(1 + \sqrt{1 - \frac{4 \cdot 4.8}{30}}\right) = 24\,s$$

$$T'_d = \frac{T_i}{2}\left(1 - \sqrt{1 - \frac{4T_d}{T_i}}\right) = \frac{30}{2}\left(1 - \sqrt{1 - \frac{4 \cdot 4.8}{30}}\right) = 6\,s$$

3.5

$$T_i = T'_i + T'_d = T'_i + T'_i/4 = 1.25\,T'_i$$

$$T_d = \frac{T'_i T'_d}{T'_i + T'_d} = \frac{T'_i T'_i/4}{T'_i + T'_i/4} = \frac{T'_i/4}{1.25} = 0.2\,T'_i$$

It follows that $T_i/T_d = 1.25/0.2 = 6.25$.

3.6 Because

$$1 - \frac{4T_d}{T_i} = 0$$

we have $T_d = T_i/4$ and the integral time will be $T'_i = T'_d = T_i/2$. This means that the integral time and the derivative time are the same and the ratio between them is 1.

3.7 If the controller has a setpoint weight that is greater than zero, you can reduce the setpoint weight. This makes it possible to tune the controller more aggressively without getting an overshoot during setpoint changes.

3.8

- Change in answer 1) corresponds to Figure E-3.1c. Worse damping because the D-part was removed.

- Change in answer 2) corresponds to Figure E-3.1d. Stationary control error (offset) because of the removal of the I-part.

- Change in answer 3) corresponds to Figure E-3.1b. The step change in the controller output for setpoint changes has disappeared.

3.9 Section 3.5 describes how to determine PV_{range}. Because the controller works with normalised signals the following equation holds:

$$\frac{\Delta u}{OP_{range}} = K \frac{\Delta e}{PV_{range}}$$

Thus, the measurement range can be determined as

$$PV_{range} = K \frac{\Delta e}{\Delta u} OP_{range} = 0.5 \frac{100}{10} 100 = 500\,^\circ C$$

3.10 We see in Figure E-3.2 that the setpoint changes from $10\,\text{m}^3/\text{h}$ to $14\,\text{m}^3/\text{h}$, which means that the step change in the control error is $\Delta e = 14 - 10 = 4\,\text{m}^3/\text{h}$. This step in the control error results in a step in the controller output with size $\Delta u = 80{-}50 = 30\,\%$. From the solution of Problem 3.9 we find that PV_{range} becomes

$$PV_{range} = K \frac{\Delta e}{\Delta u} OP_{range} = 1.5 \frac{4}{30} 100 = 20\,\text{m}^3/\text{h}$$

4. PID-Controller Tuning

4.1 Figure E-4.1a shows a case of control where the process variable slowly creeps back towards the setpoint due to an excessively long integral time. The integral time should therefore in this case be reduced, that is, the integral action should be increased. Figure E-4.1b shows an oscillating controller caused by an excessively high controller gain. The gain should therefore be reduced.

4.2

- Parameter settings 1 corresponds to the response in Figure E-4.2c. The higher gain causes the oscillations in the response.

- Parameter settings 2 corresponds to the response in Figure E-4.2b. The longer integral time gives a slower response.

- Parameter settings 3 corresponds to the response in Figure E-4.2d. The higher gain and the longer integral time gives a slower and oscillating response.

4.3 From the solution in Problem 2.4 we get the parameters $K_p = 2$, $L = 5$ s and $T = 15$ s. This gives the process time $T_p = L + 3T = 50$ s. The equations in Table 4.7 give the following controller settings for the three methods:

AMIGO:	$K = 0.32$	$T_i = 13$ s
Lambda:	$K = 0.38$	$T_i = 15$ s
One-third rule:	$K = 0.17$	$T_i = 17$ s

The AMIGO method and the Lambda method give similar results while the One-third rule has about half the controller gain compared to the other methods and approximately the same integral time.

4.4 The solutions of Problem 2.5 give the parameters $K_p = 3$, $L = 13$ s and $T = 19$ s. The equations in Table 4.7 give the following controller parameters for the different methods:

AMIGO-PI:	$K = 0.10$	$T_i = 18$ s	
Lambda-PI:	$K = 0.20$	$T_i = 19$ s	
AMIGO-PID:	$K = 0.29$	$T_i = 18$ s	$T_d = 5.4$ s
Lambda-PID:	$K = 0.33$	$T_i = 26$ s	$T_d = 4.8$ s

The PI-controller integral times are approximately the same but the Lambda method has double the gain compared to the AMIGO method. The PID-controller have similar settings but the Lambda method has a somewhat longer integral time.

4.5 From the solutions of Problem 2.6 we get the parameters $K_v = 0.1$ s^{-1} and $L = 5$ s. The equations in Table 4.8 give the following controller parameters for the different methods:

Ziegler-Nichols-PI:	$K = 1.8$	$T_i = 15$ s	
AMIGO-PI:	$K = 0.70$	$T_i = 67$ s	
Ziegler-Nichols-PID:	$K = 2.4$	$T_i = 10$ s	$T_d = 2.5$ s
AMIGO-PID:	$K = 0.90$	$T_i = 40$ s	$T_d = 2.5$ s

There are big differences between the methods. The Ziegler-Nichols method provides significantly more aggressive control than the AMIGO method with higher gains and shorter integral times.

4.6 The solution to Problem 2.9 gives the parameters $K_p = -2.3$, $L = 45$ s and $T = 37$ s. The fact that the process gain is negative is taken into account by selecting *direct* control instead of *reverse*. In the calculation of the controller parameters, the absolute value $K_p = 2.3$ is therefore used. The equations in Table 4.7 give the following controller parameters for the different methods:

AMIGO-PI:	$K = 0.10$	$T_i = 38$ s	
Lambda-PI:	$K = 0.20$	$T_i = 37$ s	
AMIGO-PID:	$K = 0.25$	$T_i = 44$ s	$T_d = 16$ s
Lambda-PID:	$K = 0.43$	$T_i = 60$ s	$T_d = 14$ s

The PI-controllers have essentially the same integral times, but the Lambda method has twice the gain of the AMIGO method. For the PID-controllers, the AMIGO method gives a lower gain and a shorter integral time than the Lambda method, while the derivative times are approximately the same. The methods suggest higher gain for the PID-controllers than for the PI-controllers.

4.7 The solution to Problem 2.10 gives the parameters $K_p = -2.2$, $L = 40\,\text{s}$ and $T = 10\,\text{s}$. The fact that the process gain is negative is taken care of in the controller by choosing *direct* control instead of *reverse*. In the calculation of the controller parameters, the absolute value $K_p = 2.2$ is therefore used. The process time can be calculated to $T_p = L + 3T = 70\,\text{s}$, which fits well with what can be seen in Figure E-2.8. The step response starts at time $t = 80\,\text{s}$ and the process variable has reached its new level approximately at time $t = 150\,\text{s}$. The equations in Table 4.7 give the following controller parameters for the three methods:

AMIGO:	$K = 0.090$	$T_i = 17\,\text{s}$
Lambda:	$K = 0.028$	$T_i = 10\,\text{s}$
One-third Rule:	$K = 0.15$	$T_i = 23\,\text{s}$

The AMIGO method and the One-third rule give settings that are quite close to each other, while the Lambda method has a much lower gain and shorter integral time.

4.8 From the solution to Problem 2.10 we can calculate the process parameters $K_p = (0.14 + 0.12)/2 = 0.13$, $L = 0\,\text{s}$ and $T = (400 + 300)/2 = 350\,\text{s}$. Both the Ziegler-Nichols method and the AMIGO method give infinitely large controller gains and the integral time $T_i = 0$. The only methods that work are therefore the Lambda method and the One-third rule. The formulas in Table 4.7 give the following controller parameters for these two methods:

Lambda:	$K = 7.7$	$T_i = 350\,\text{s}$
One-third rule:	$K = 2.6$	$T_i = 350\,\text{s}$

For the Lambda method, λ is chosen to be $\lambda = T$ and for the One-third rule, T_p has been calculated with the formula $T_p = L + 3T$. Both the Lambda method and the One-third rule give the integral time $T_i = T$. The Lambda method gives a gain that is three times greater than the One-third rule.

4.9 From the solution of Problem 2.12 we get the process parameter $K_p = 0.75$, $L = 2\,\text{s}$ and $T = 6\,\text{s}$. The process time is $T_p = 20\,\text{s}$. The equations in Table 4.7 give us the following controller parameter for the four methods:

Ziegler-Nichols:	$K = 3.6$	$T_i = 6\,\text{s}$
AMIGO:	$K = 0.85$	$T_i = 5.2\,\text{s}$
Lambda:	$K = 1.0$	$T_i = 6\,\text{s}$
One-third rule:	$K = 0.44$	$T_i = 6.7\,\text{s}$

The different methods have approximately the same integral times. The Ziegler-Nichols method gives a significantly higher gain than the other methods and the One-third rule gives a gain that is about half as large as the one calculated with the AMIGO method and the Lambda method.

4.10   If you choose $T = 0$ in the expressions in Table 4.7 to compute the tuning parameters you get:

AMIGO-PI:	$K = 0.15/K_p$	$T_i = 0.35L$	
Lambda-PI:	$K = 0$	$T_i = 0$	
AMIGO-PID:	$K = 0.2/K_p$	$T_i = 0.4L$	$T_d = 0$
Lambda-PID:	$K = 0.14/K_p$	$T_i = 0.5L$	$T_d = 0$

Both the AMIGO method and the Lambda method suggest PI-controllers because $T_d = 0$ in the equations of the PID-controllers. The Lambda method for PI-controllers does not work, because it suggests that $K = T_i = 0$. The other three tuning methods, AMIGO-PI, AMIGO-PID and Lambda-PID, suggest parameters relatively close to one another.

4.11   The step response will be completely dominated by the long dead time, and the time constant can be neglected. This means that $T_p \approx L$. The One-third rule therefore gives $K = K_p/3$ and $T_i = L/3$. This setting is close to the settings given by the AMIGO method and the Lambda method in Problem 4.10, but with a slightly higher gain.

4.12   Ziegler-Nichols method gives: $K = 0.4K_c = 2.4$, $T_i = 0.8T_c = 16\,\text{s}$
AMIGO method gives: $K = 0.16K_c = 0.96$, $T_i = \frac{K_pK_c}{K_pK_c+4.5}T_c = 15\,\text{s}$

The integral times are approximately equally long but the Ziegler-Nichols method gives a much higher controller gain than the AMIGO method.

5.   Nonlinear Processes

5.1   Figure S-5.1 shows the valve characteristic. The points on the curve mark the measured values of process variable y plotted against the corresponding values of controller output u when the process variable is stationary at different levels.

5.2   Since the valve causes the nonlinearity, it is appropriate to choose the controller output u as the reference variable GS_{ref}. The valve is quick-opening, which means that the gain changes most at small valve openings and least at large valve openings. On the assumption that you want good control over the entire operating range, the limits should therefore be closer together at small valve openings. The gain becomes very high at extremely small valve openings, so it is not realistic to be able to control the process in this area at all. Reasonable choices are to put the first limit at about $u = 10\,\%$ and the second at about $u = 30\,\%$.

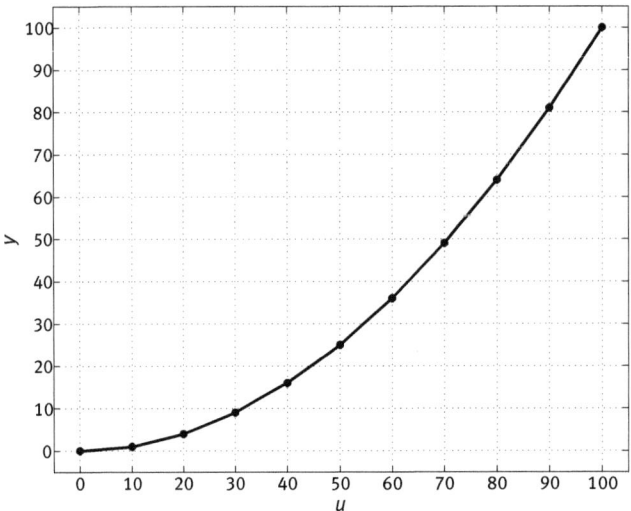

Fig. S-5.1: Valve characteristic in Problem 5.1.

5.3 The step change at time $t = 10$ gives a response in the process variable. This means that the step made in the same direction at time $t = 50$ is made beyond the backlash. Here, a change in the controller output of 20 % gives a change in the process variable of 20 %. This means that the process gain is $K_p = 1$. During the downward step change at time $t = 90$, the process variable only comes down to 45 % instead of the 50 % it was at when the backlash was at the end of the upward phase. This means that the gap in the process variable is $\Delta y = 5$ %. Since $K_p = 1$, it means that the gap in the controller output is also $\Delta u = 5$ %.

5.4 The oscillation is caused by valve stiction. The flow measurement has the characteristic appearance similar to a square wave and the controller output resembles a triangular wave.

5.5 The slow control is due to the controller being tuned conservatively with either too low a gain or too long an integral time, or both. The oscillations in the signals are due to high friction in the valve. When the valve "disengages", the control error suddenly decreases and thus also the proportional part in the controller output. This causes the controller output to reverse.

5.6 All changes in the controller output have a magnitude of 2 %. In the ascending series of step changes, we get a response in the process variable at all changes except the first. We have therefore definitely exceeded the backlash when the controller output is at levels $u = 19, 21, 23$ and 25 %. At the descending sequence, we also get responses to all changes except the first, which means that we are guaranteed to have exceeded the backlash level when the controller output is at levels u = 21, 19, 17, 15 and 13 %. The process variable is at the same level when the controller output is 19 % during the increase as when it is 17 % during the decrease. The same applies to the levels 21 % during the increase and 19 % during the decrease. From this we can conclude that the size of the gap is 2 %. The changes in the process variable are greater with small controller outputs than with large ones. This shows that we have a quick-opening valve.

5.7 The large variation in the controller output that does not affect the process variable can be due to backlash or friction. It is difficult to determine whether the problem here is backlash or friction, but the sharp change at the time around 7000 seconds indicates that there is at least some friction that has caused excessive pressure to build up in the actuator causing a stick-slip movement.

5.8 There is significant backlash in the valve because of the following argument. The setpoint change looks good and the gap is closed. The process variable settles on the setpoint which is close to 70 %. At about 200 seconds, something happens that causes the process variable to end up a bit above the setpoint, which causes the controller to increase the controller output. However, nothing happens to the process variable during the time of the upset, but the controller output only drifts within the gap.

5.9 There are many ways to measure the backlash in the figure. One way is as follows. Study the conditions when the controller output is at 27 %. We have two places where this happens in an upward phase with the gap closed, namely at times 80 s and 680 s. The process variable here is approximately at 62.5 %. We have a measurement in the downward phase when the gap is closed, namely at 360 s. Here the process variable is approximately 59 %. We thus have a gap in the process variable which is Δy = 62.5 − 59 = 3.5 %. To determine the backlash in the controller output, we have to divide by the process gain. At, for example, time 90 s, we have a step response where the controller output goes from 27 % to 29.5 % and the process variable from 62.5 % to 57.5 %. The process gain therefore can be calculated as K_p = (57.5−62.5)/(29.5−27) = −2. That gives us the controller output gap $\Delta u = \Delta y / |K_p|$ = 1.75 ≈ 2 %.

6. Control Strategies

6.1 In subplot b, the low-frequency signal has been almost completely filtered out, while the high-frequency signal has passed through. In subplot f, the low-frequency component is basically completely gone and the amplitude of the high-frequency component has been reduced. These partial figures correspond to the two high-pass filters, where T_{hp} = 1 s in b and T_{hp} = 0.1 s in f. Subplots c, d and e correspond to low-pass filters where d has such a long filter time constant that the amplitude of the low-frequency component has decreased significantly. Subplot d corresponds to $T_{lp} = 20$ s. In subplot e, the high-frequency component has a higher amplitude than in subplot c, so subplot e corresponds to $T_{lp} = 0.5$ s and c to $T_{lp} = 1$ s.

6.2 The filter time constant must be longer than 1 s in order for us to get a noticeable reduction of the disturbance signal. However, the filter time constant must not be so long that it significantly affects the dynamics of the process. As the integral time is chosen to 30 s, it is likely that the process has a dynamic that is close to 30 s. A suitable choice is therefore around 5 to 10 s.

6.3 T_{hp} should be chosen equal to the time constant of the original filter, that is, $T_{hp} = 30$ s. T_{lp} should be chosen as the time constant of the desired filter, i.e. $T_{lp} = 5$ s. For explanation, see Example 6.3 in Section 6.2.

6.4 a. 40 %, b. 10 %, c. A is a MAX-selector and C is a MIN-selector. B can be both a MAX- and a MIN-selector. d. B is a MIN-selector and C is a MAX-selector. A can be both a MAX- and a MIN-selector.

6.5
$$K_f = \frac{u_1 - u_2}{v_1 - v_2} = \frac{30 - 20}{45 - 50} = -2.$$

6.6 When the load disturbance changes from 20 % to 40 % the controller output need to change from 50 % to 40 %. This gives the following feed forward gain:
$$K_f = \frac{u_1 - u_2}{v_1 - v_2} = \frac{50 - 40}{20 - 40} = -0.5.$$

6.7 If you supply the temperature signal to the flow controller and there is a sudden increase in the water temperature, the following will happen. The increase will mean that the flow controller reduces the valve position, which in turn means that the steam flow on the primary side decreases. So long, everything is fine. However, as the flow decreases, the flow controller will react to the fact that the flow has become lower than the setpoint, which in turn means that the controller opens the valve again to eliminate the control error. We thus get no lasting reduction in steam flow. The correct action is to supply the temperature signal to the temperature controller. In the event of a change in the input water temperature, the feed-forward will then mean a reduction in the setpoint to the flow controller.

6.8 Split-range control. The excess gas is used for smaller controller output levels, say up to 50 %. At 50 % the valve of the excess gas is fully open. The natural gas is used between 50 % and 100 %. The natural gas valve is closed at a controller output signal of 50 % and fully open at 100 %. The range limit of the split-range-controller at 50 % can be adjusted so that a similar dynamic results in both cases. Another way to achieve this is to use gain scheduling.

Bibliography

[1]   W. Altmann, D. Macdonald, and S. Mackay. *Process Control for Engineers and Practitioners*. Elsevier, Amsterdam, The Netherlands, 2005.

[2]   A. Andersson. *Measurement Technology for Process Automation*. CRC Press, 2017.

[3]   K. Åström and T. Hägglund. *PID Controllers: Theory, Design, and Tuning*. ISA - The Instrumentation, Systems and Automation Society, Research Triangle Park, North Carolina, 1995.

[4]   K. Åström and T. Hägglund. *Advanced PID Control*. ISA - The Instrumentation, Systems and Automation Society, Research Triangle Park, North Carolina, 2006.

[5]   B. W. Bequette. *Process Control: Modeling, Design and Simulation*. Pearson, Research Triangle Park, NC, 2002.

[6]   T. L. Blevins, G. K. McMillan, W. K. Wojsznis, and M. W. Brown. *Advanced Control Unleashed*. ISA, Research Triangle Park, NC, 2003.

[7]   C. Kravaris and I. K. Kookos. *Understanding Process Dynamics and Control*. Cambridge University Press, Cambridge, UK, 2021.

[8]   S. E. LeBlanc and D. R. Coughanowr. *Process Systems Analysis and Control*. McGraw-Hill, New York, third edition edition, 2009.

[9]   B. G. Lipták. *Process Control: Instrument Engineers' Handbook*. Butterworth-Heinemann, 2013.

[10] M. L. Luyben and W. L. Luyben. *Essentials of Process Control*. McGraw-Hill, London, UK, 1997.

[11] T. E. Marlin. *Process Control*. McGraw-Hill, 2000.

[12] A. O'Dwyer. *Handbook of PI and PID Controller Tuning Rules*. Imperial College Press, London, UK, 2009.

[13] B. A. Ogunnaike and W. H. Ray. *Process Dynamics, Modeling and Control*. Oxford University Press, Oxford, UK, 1994.

[14] D. E. Seborg, T. F. Edgar, and D. A. Mellichamp. *Process Dynamics and Control*. Wiley, New York, second edition edition, 2004.

[15] F. G. Shinskey. *Controlling Multivariable Processes*. ISA, Research Triangle Park, North Carolina, 1981.

[16] F. G. Shinskey. *Feedback Controllers for the Process Industries*. McGraw-Hill, New York, 1994.

[17] F. G. Shinskey. *Process-Control Systems. Application, Design, and Tuning*. McGraw-Hill, New York, fourth edition edition, 1996.

[18] S. Skogestad and I. Postlethwaite. *Multivariable Feedback Control*. Wiley, second edition edition, 2005.

[19] SSG Skogsindustriernas Teknik AB. *Regleroptimering*. Rapport SSG 5253, Sundsvall, 2004.

[20] G. Stephanopoulos. *Chemical Process Control: An introduction to theory and practice*. Prentice Hall, New Jersey, 1984.

[21] A. Visioli. *Practical PID Control*. Springer, Berlin, 2006.

[22] H. Wade. *Basic and Advanced Regulatory Control: System Design and Application*. ISA, Research Triangle Park, NC, 2004.

https://doi.org/10.1515/9783111104959-009

Index

adaptive control 97
adaptive feedforward 126
air-fuel control 142
AMIGO-method
– self-oscillating method 76
– step response method 59
anti-windup 48
asymmetrical processes 92
auto-tuning 77
automatic tuning 77

backlash 88
balance tanks 154
batch manufacturing 154
batch reactor 154
big-valve-little-valve 127
boiler level control 120
buffer control 154
buffer tank 154

cascade control 110
– commissioning 116
– integral windup 116
– tracking 116
– tuning 115
concentration control 91, 118, 135
control strategies 101
controller gain 7
– identification 38
controller implementation 49
controller output range 7
controller structure
– identification 40
critical gain 72
critical time period 72

D-part 26
damped process 9
damping 6
dead time 6, 135
– step response example 18
dead time compensation 135
dead time process 10
– occurances 135
dead zone 102, 129, 132, 155
decoupling 150

derivative time
– identification 40
differential pressure sensor 90
disturbance filtering 101
disturbances 52, 101
dither signal 87
DPID-controller 28

excessive quantisation 102
external control mode 115

feedforward 117
– concentration control 118
– lead/lag filter 129
– practical realisation 124
– with cascade control 125
– with mid-range control 129, 131
– adaptive 126
feedforward mid-range control 131
FFMRC 131
filter 46
– of derivative part 46
– PID-controller 46
filtering
– dead zone 102
filters 101
– high-pass filter 104
– lead/lag filter 106
– low-pass filter 103
first order process 9
flow measurement 90
friction 83
fuel-air control 142

gain scheduling 92

heat exchanger control
– cascade control 110, 114
– gain scheduling 95
heating control 121
high-pass filter 104
higher order process 9
– step response analysis 16
hunting 84
hysteresis 88

I-part 24

https://doi.org/10.1515/9783111104959-010

implementation 49
integral time
– identification 39
integral windup 48
– with cascade control 116
– in selector control 110
integrating controller 24
integrating process 10
– step response analysis 14, 17
interacting form 30
internal control mode 115

knocker 87

lag time 6
Lambda method 61
lead/lag filter 106
level measurement 90
limitations 48
linearisation 92
linearisation blocks 83
load disturbances 52
low-pass filter 46, 103

manual tuning 54
master controller 112
material transport 10
measurement noise 46, 52, 101
measurement range 7
mid-range control 126
MIN/MAX-selector 108

network 136
neutralisation 91, 118, 135
noise 46, 52, 101
noise filtering 101
non-interacting form 30
nonlinear processes 81
nonlinear sensors 89

on/off controller 20
– selection 31
one-element control 120
one-third tuning rule 62
oscillating process 10
oscillation 85
oscillation diagnosis 85
Otto-Smith controller 138
override control 108

overshoot 6

P-controller
– selection 32
P-part 21
parallel form 29
– identification 41
PD-controller
– selection 32
pH-control 91, 118, 135
PI-controller
– selection 33
PID-controller 20
– identification 38
– tuning 52
– dead time processes 138
– ideal form 29
– modification 43
– parallel 29
– parameters 34
– selection 33
– series 29
– structure 20
positioner 85
PPI-controller 141
– tuning 141
primary controller 112
process 8
process disturbance 52
process gain 6
process models 5
process time 63
process types 5
process variable range 7
– identification 42
proportional band 35
Pt100-sensor 90

quantisation 46, 102

ratio control 142
ratio station 142
recycle 10
relay method 72
– auto-tuning 79
repeats per minute 37
repeats per second 37
reset time 24
resistance thermometer 90

reverse response 11
robustness 6, 53
rpm 37
rps 37

sampled controllers 136
sampling period 49
secondary controller 112
selector 108
self-oscillation method 70
– relay method 72
– Ziegler-Nichols method 72
series form 29
– identification 41
setpoint 2
– weighting 44
sixty-three percent method 12
slave controller 112
speed gain 15
– example 17
split-range control 132
steady-state gain 6
steam boiler 11
step response 5
step response analysis 12
– auto-tuning 78
stick-slip motion 84
stiction 83

temperature sensor 89
termocouple 90
three-element control 120
time constant 6
time delay 6
tracking 110
– with cascade control 116
tracking ratio station 148
transport delay 136
TRS 148
tuning diagrams 55
tuning methods 52

valve 81
– backlash 88
– friction 83
– characteristic 81
valve position controller – VPC 127
valve positioner 85, 115
velocity gain 15
volume measurement 90
VPC – valve position controller 127

Ziegler-Nichols method
– self-oscillation method 72, 75
– step response method 57